T0292121

These notes arise from lectures presented in Florence under the auspices of the Accademia Nazionale dei Lincei and deal with an area that lies at the crossroads of mathematics and physics. The material presented here rests primarily on the pioneering work of Vaughan Jones and Edward Witten relating polynomial invariants of knots to a topological quantum field theory in $2+1$ dimensions. Professor Atiyah here presents an introduction to Witten's ideas from the mathematical point of view. The book will be essential reading for all geometers and gauge theorists as an exposition of new and interesting ideas in a rapidly developing area.

The geometry and
physics of knots

Lezioni Lincee
Editor: Luigi A. Radicati di Brozolo, Scuola Normale Superiore, Pisa

This series of books arises from a series of lectures given under the auspices of the Accademia Nazionale dei Lincei through a grant from IBM.

The lectures, given by international authorities, will range on scientific topics from mathematics and physics through to biology and economics. The books are intended for a broad audience of graduate students and faculty members, and are meant to provide a '*mise au point*' for the subject they deal with.

The symbol of the Accademia, the lynx, is noted for its sharp sightedness; the volumes in the series will be penetrating studies of scientific topics of contemporary interest.

Already published

Chaotic Evolution and Strange Attractors: D. Ruelle
Introduction to Polymer Dynamics: P. de Gennes

The geometry and physics of knots

MICHAEL ATIYAH
Master of Trinity College, Cambridge

CAMBRIDGE
UNIVERSITY PRESS

Published by the Press Syndicate of the University of Cambridge
The Pitt Building, Trumpington Street, Cambridge CB2 1RP
40 West 20th Street, New York, NY 10011, USA
10 Stamford Road, Oakleigh, Melbourne 3166, Australia

First published 1990
Reprinted 1991, 1993

British Library cataloguing in publication data available

Library of Congress cataloguing in publication data available

ISBN 0 521 39521 6 hardback
ISBN 0 521 39554 2 paperback

Transferred to digital printing 2004

AS

Contents

Preface

These lecture notes are an expanded version of the series of lectures I gave, by invitation of the Accademia Nazionale dei Lincei, at the University of Florence in November 1988. They have also benefited from the seminar I ran in Oxford during that Autumn term. I am grateful in particular to Graeme Segal, Nigel Hitchin and Ruth Lawrence who helped me to run that seminar and to clarify many of the issues involved. I would also like to thank the mathematicians in Florence for providing such a receptive audience.

Sometimes a series of lectures may be the culmination of many years of work on a topic. In that case lecture notes may take on a definitive form, providing a careful treatment of the subject. On other occasions the lectures may come at the beginning of some new development, in which case they provide an introduction to current and future work. This is the case with these present lecture notes. The subject they deal with is just opening up and is now developing at a rapid rate. Moreover the area lies at the crossroads of mathematics and physics. This adds greatly to its interest but increases the difficulty of presentation. In due course a coherent and polished mathematical account will emerge but these lecture notes make no pretence to fulfill that role.

I have to a great extent followed the lines of the lectures as they were delivered. This means I have emphasized motivation and ideas at the expense of technicalities and formulae. As a result the reader will find no theorems even formulated, let alone proved, in the text. However, I have provided an extensive list of references where many of the relevant details can be found.

The material presented here rests primarily on the pioneering ideas of Vaughan Jones and Edward Witten. I have benefited greatly from extensive discussions with both of them and I hope these notes may serve a useful purpose by introducing their magnificent ideas to a wide mathematical public.

Oxford, September 1989

1

History and background

1.1 General introduction

In recent years there has been a remarkable renaissance in the interaction between geometry and physics. After a long fallow period in which mathematicians and physicists pursued apparently independent paths their interests have now converged in a striking manner. However, it appears that parallel problems were being investigated in the past but a common language and framework were missing. This has now been rectified with gauge theory (alias the theory of connections) providing the common ground.

In earlier periods geometry and physics interacted at the classical level, as in Einstein's theory of general relativity, with gravitational force being interpreted in terms of curvature. The new feature of the present interaction is that quantum theory is now involved and it turns out to have significant relations with topology. Thus geometry is involved in a global and not purely local way.

A somewhat surprising feature of the new developments is that quantum field theory seems to tie up with deep properties of low-dimensional geometry, i.e. in dimensions 2, 3 and 4 [3]. Thus the exciting new results of Donaldson [10] on four-dimensional manifolds, and the associated theory of Floer [13] on three-dimensional manifolds, are intimately linked to Yang–Mills theory. This has been made even clearer by Witten [35], where the Donaldson–Floer theory is interpreted as a *topological* quantum field theory in $3+1$ dimensions.

A slightly different case arises from the recently discovered polynomial invariants of knots by Vaughan Jones [17]. These are related to physics in various ways but the most fundamental is due to Witten [36] who has shown that the Jones invariants have a natural interpretation in terms of a topological quantum field theory in $2+1$ dimensions. My purpose in these lectures is to present this new theory of Witten. Shortage of time and the present novelty and incompleteness of the theory mean that this is not a definitive treatment. Rather it is an introduction to Witten's ideas, presented from the mathematical point of view. The whole subject is still developing rapidly and a provisional account accessible to mathematicians may serve a useful purpose.

1.2   Gauge theories

The prototype of all gauge theories is electromagnetism. From the geometrical point of view the electromagnetic potential a_μ ($\mu = 1, \ldots, 4$) defines a connection for a $U(1)$ bundle over Minkowski space M. The field is the corresponding curvature

$$f_{\mu\nu} = \partial_\mu a_\nu - \partial_\nu a_\mu \quad (\partial_\mu = \partial/\partial x_\mu).$$

Maxwell's equations in vacuo take the form

$$\mathrm{d}f = 0, \quad \mathrm{d}^* f = 0$$

where f is now viewed as a 2-form, d is the exterior derivative and d^* is its formal adjoint (relative to the Minkowski metric).

Non-abelian gauge theories are obtained by replacing $U(1)$ with a compact non-abelian Lie group G, e.g. $SU(n)$. A potential is then a connection A over Minkowski space, with components A_μ in the Lie algebra of G, and the field is the curvature F with components

$$F_{\mu\nu} = \partial_\mu A_\nu - \partial_\nu A_\mu + [A_\mu, A_\nu].$$

The most straightforward generalization of Maxwell's

equations are the Yang–Mills equations†

$$\mathrm{d}F = 0, \quad \mathrm{d}^*F = 0,$$

where d and d* are *covariant* derivatives. Gauge theories possess an infinite-dimensional symmetry group given by functions $g: M \to G$ and all physical, or geometric, properties are gauge invariant.

To specify a physical theory the usual procedure is to define a Lagrangian or action L. This is a functional of the various fields obtained by integrating over M a Lagrangian density. For example, for a scalar field theory where the only field is a scalar function φ, the simplest Lagrangian is

$$L(\varphi) = \int_M |\mathrm{grad}\ \varphi|^2 \,\mathrm{d}x$$

where the norm and volume are those of Minkowski space.

For Yang–Mills theory the Lagrangian is

$$L(A) = \int_M |F_A|^2 \,\mathrm{d}x$$

where the norm here also uses an invariant metric on G.

Having fixed a Lagrangian $L(\varphi)$ the 'partition function' of the theory (by analogy with statistical mechanics) is the Feynman functional integral

$$Z = \int \exp{(\mathrm{i}L)} \,\mathrm{D}\varphi.$$

More generally, for any functional $W(\varphi)$, the unnormalized expectation value of the 'observable' W is defined by the integral

$$\langle W \rangle = \int \exp{(\mathrm{i}L(\varphi))}\, W(\varphi)\, \mathrm{D}\varphi.$$

These Feynman integrals are not very well defined mathematically but they can, when used skilfully, be a useful heuristic

† Strictly $\mathrm{d}^*F = 0$ is the Yang–Mills equation and $\mathrm{d}F = 0$ is the Bianchi identity.

tool. In particular, perturbation expansions can be computed explicitly.

The Feynman integral provides a relativistically invariant approach. This is its main purpose. In a non-relativistic treatment a quantum field theory is described by a time-evolution operator e^{itH} in a certain Hilbert space $\mathcal{H}$. The infinitesimal generator H is the Hamiltonian of the theory. There are formal rules which, starting from the Lagrangian formulation via the Feynman integral, produce the Hilbert space $\mathcal{H}$ and the Hamiltonian H. The fundamental relation between the two approaches rests on the formula

$$\langle \exp{(iTH)}\varphi_0, \varphi_T \rangle = \int \exp{(iL(\varphi))} \, D\varphi$$

where φ_0, φ_T are scalar fields on R^3 (space) and the Feynman integral is taken over all fields $\varphi(x, t)$ which interpolate between $\varphi_0 = \varphi(x, 0)$ and $\varphi_T = \varphi(x, T)$ for $0 \le t \le T$. In particular

$$\text{Trace} \exp{(iTH)} = \int \exp{(iL(\varphi))} \, D\varphi \qquad (1.2.1)$$

where, in the Feynman integral, φ is a function on $R^3 \times S^1_T$ where S^1_T is the circle of length T.

Witten's version of the Jones theory is defined by a suitable choice of Lagrangian in $2+1$ dimensions and this will be described in Chapter 7. Until then we shall be following the non-relativistic Hamiltonian approach, which is mathematically more rigorous.

In gauge theory, classical fields of force are described in terms of curvature. However, gauge theories have global features which can be non-trivial even when all curvatures vanish. This is fundamental for the relations with quantum field theory which are our basic interest. The prototype of this is the Bohm–Aharonov effect in the quantum theory of the electron. This concerns a solenoid with an interior magnetic flux but with no external magnetic field. A beam of electrons travelling past the solenoid produces interference patterns indicating a phase-shift. This physical effect takes place even though the electrons travel in a force-free region.

Mathematically the wave-function of the electron in the external region is a section of a flat line-bundle, with non-trivial holonomy round the solenoid.

In non-abelian gauge theories wave-functions are sections of vector bundles and the holonomy lies in a non-abelian group. This is the starting point for the relation between topology and quantum field theory that is embodied in the Jones–Witten theory.

1.3   History of knot theory

The study of knots (and links) in ordinary three-dimensional space is the archetype of a topological problem. Knots are remarkably complicated things and, even with all the sophisticated techniques of modern topology, they have resisted a definitive treatment. The remarkable developments growing out of the Jones polynomial are an indication of the subtlety of knot theory.

A *knot* is by definition a smooth embedding of a circle in R^3. Two knots are equivalent if one knot can be deformed continuously into the other without crossing itself. A *link* is an embedded finite union of disjoint circles.

Knot theory has an interesting history. In the nineteenth century physicists were pondering on the nature of atoms. Lord Kelvin, one of the leading physicists of his time, put forward in 1867 the imaginative and ambitious idea that atoms were knotted vortex tubes of ether [32].

The arguments in favour of this idea may be summarized as follows.

(1) *Stability*. The stability of matter might be explained by the stability of knots (i.e. their topological nature).

(2) *Variety*. The variety of chemical elements could be accounted for by the variety of different knots.

(3) *Spectrum*. Vibrational oscillations of the vortex tubes might explain the spectral lines of atoms.

From a modern twentieth-century point of view we could, in retrospect, have added a fourth.

(4) *Transmutation*. The ability of atoms to change into other atoms at high energies could be related to cutting and recombination of knots.

For about 20 years Kelvin's theory of vortex atoms was taken seriously. Maxwell's verdict was that 'it satisfies more of the conditions than any atom hitherto considered'.

Kelvin's collaborator P. G. Tait undertook an extensive study and classification of knots [31]. He enumerated knots in terms of the crossing number of a plane projection and also made some pragmatic discoveries which have since been christened 'Tait's conjectures'. After Kelvin's theory was discarded as an atomic theory the study of knots became an esoteric branch of pure mathematics.

Despite the great strides made by topologists in the twentieth century the Tait conjectures resisted all attempts to prove them until the late 1980s. The new Jones invariants turned out to be powerful enough to dispose of most of the conjectures fairly quickly.

One of the early achievements of modern topology was the discovery in 1928 of the *Alexander polynomial* of a knot or a link [1]. Although it did not help to prove the Tait conjectures it was an extremely useful knot invariant and greatly simplified the effective classification of knots. The Alexander polynomial arises from the homology of the infinite cyclic cover of the complement of a knot. Equivalently it can be derived from considering cohomology of the knot complement with coefficients in a flat line-bundle. This is very much the context of the Bohm–Aharonov effect.

For more than 50 years the Alexander polynomial remained the only knot invariant of its kind. It was therefore a great surprise to all the experts when, in 1984, Vaughan Jones discovered another polynomial invariant of knots and links. As already mentioned, this turned out to be extremely useful and enabled several of Tait's conjectures to be established.

In the next section we shall briefly summarize some of the key facts about the Jones polynomials. For an excellent and thorough presentation the reader is referred to the account by Jones in [17].

1.4   The Jones polynomial

The Jones polynomial is a polynomial in t, t^{-1} assigned to a knot K in R^3. It is denoted by $V_K(t)$. It is normalized so that $V(t) = 1$ for the unknot (the standard unknotted circle in R^3). Moreover it has the key property

$$V_{K^*}(t) = V_K(t^{-1}) \qquad (1.4.1)$$

where K^* is the mirror image of K. Simple examples show that $V_K(t)$ need not be invariant under $t \to t^{-1}$, so that the Jones polynomial can sometimes distinguish knots from their mirror images. For example the right-handed trefoil knot has

$$V(t) = t + t^3 - t^4$$

and so is distinguished from its mirror image. The Alexander polynomial on the other hand always takes the same value for a knot and its mirror image.

The Jones polynomial can be defined (as a Laurent polynomial in $t^{1/2}$) generally, for any *oriented link L* (i.e. each component of L is oriented). Reversing the orientation of all components leaves the Jones polynomial unchanged. This explains why, for a knot, the orientation is irrelevant.

If we represent a link by a general plane projection with over/under crossings the Jones polynomial can be characterized and computed by a skein relation. Given any oriented link diagram L, and a crossing point, we can alter the crossing to produce three different diagrams as indicated

Let V_+, V_-, V_0 denote the Jones polynomials of these links. Then the skein relation is

$$t^{-1}V_+ - tV_- = (t^{1/2} - t^{-1/2})V_0. \qquad (1.4.2)$$

The skein relation is, in a sense, deceptively simple. There is no obvious reason *a priori* why this relation should define a link invariant: it might depend on the plane presentation.

The way the Jones polynomial was originally discovered was via braids and representations of the Hecke algebra. A braid is a collection of strands as depicted below.

Note that all strands move upwards. Two braids can be composed in an obvious way, giving the braid group on n strands B_n. Formally we can define B_n as the fundamental group of the configuration space C_n of n distinct points in the plane. The usual picture of a braid can then be viewed as the space–time graph (with time vertical) of motion along a closed path in C_n.

Given a braid β we can form an oriented link $\hat{\beta}$ by closing up the braid in a standard way (see below).

Conjugate elements in B_n give rise to equivalent links. Moreover increasing the number of strands in a braid by a simple twist, as shown,

$$\biggr| \Rightarrow \biggr\rangle \hspace{3cm} (1.4.3)$$

does not affect the corresponding link. A classical theorem of Markov asserts that these two moves generate all equivalences between the resulting links.

Thus to produce an invariant of oriented links one need only produce a *class function* on all B_n which is unchanged by the move (1.4.3).

Since class functions arise naturally as characters of representations this suggests we start by considering representations of the braid groups. In fact Jones used representations which came from the *Hecke algebra* $H(n, q)$. This is the quotient of the group algebra of B_n obtained by requiring the generator σ (a single twist of consecutive strands) to satisfy the quadratic relation

$$(\sigma - q)(\sigma + 1) = 0.$$

If $q = 1$ so that $\sigma^2 = 1$ we get the group algebra of the symmetric group S_n. It follows that, for generic values of q, $H(n, q)$ has the same irreducible representations as S_n.

For each Young diagram (parametrizing an irreducible representation of S_n) we then get a character of B_n which depends on q (as a Laurent polynomial in $q^{1/2}$). The Jones polynomial (with $t = q$) is a suitable combination of these characters. In fact only the two-rowed Young diagrams are needed.

The Jones polynomial has been generalized in a variety of ways. One way, described in detail in [17], gives a two-variable polynomial. This also satisfies a skein relation and can be constructed from representations of the Hecke algebra, but now using all Young diagrams.

Another, and more fundamental, way involves choosing a compact Lie group G and an irreducible representation. A polynomial invariant of oriented links is then constructed by using solutions of the Yang–Baxter equations. The original Jones polynomial corresponds to taking $G = SU(2)$ with its standard representation on C^2. Taking $G = SU(n)$ for all n, together with their standard representation on C^n, gives polynomials which, taken together, are equivalent to the two-variable polynomial of [17].

Witten's approach, which we shall be describing, also involves a choice of group G and a representation. It produces the relevant polynomials in a more direct and natural manner. Moreover, in Witten's theory, we get invariants for links in arbitrary compact 3-manifolds alone (taking the empty link with no components). This is a major advantage and is a convincing demonstration of the naturality of Witten's method.

It is perhaps worth emphasizing that the algebraic or combinatorial definition of the Jones polynomial is quite elementary and rigorous. It lacks, however, any clear conceptual interpretation. This is precisely what Witten's theory provides, although there are still technical difficulties in developing this side of the theory in all its aspects.

While the Alexander polynomial can be understood in terms of standard algebraic topology (homology theory) and has analogues in higher dimensions, the Jones polynomial is best understood in terms of a purely three-dimensional quantum field theory. There are some indications (discussed in Chapter 6) that the quantum field theory may be related to more standard geometric constructions but this has yet to be worked out.

2

Topological quantum field theories

2.1 Axioms for a topological QFT

The notion of a topological quantum field theory (QFT), i.e. not depending on any background geometry, is one which has emerged recently in the work of Witten [35] [36]. The Jones polynomial fits into such a theory so we shall begin by reviewing briefly what is meant by a topological QFT. It is convenient to give an axiomatic approach since this emphasizes the mathematical structures involved. The physics can be viewed as motivating background.

A more extensive treatment of topological QFTs can be found in [2]. The reader may also wish to consult [3] and, for closely related ideas on conformal field theories, the treatment in [29] may be helpful.

A topological QFT in dimension d is a functor Z which assigns

(1) a finite-dimensional complex vector space $Z(\Sigma)$ to each compact oriented smooth d-dimensional manifold Σ,

(2) a vector $Z(Y) \in Z(\Sigma)$ for each compact oriented $(d+1)$-dimensional manifold Y with boundary Σ.

This functor satisfies the following axioms.

> A1 (*Involutory*) $Z(\Sigma^*) = Z(\Sigma)^*$, where Σ^* denotes Σ with the opposite orientation and $Z(\Sigma)^*$ is the dual space.
>
> A2 (*Multiplicativity*) $Z(\Sigma_1 \cup \Sigma_2) = Z(\Sigma_1) \otimes Z(\Sigma_2)$, where $\cup$ is the disjoint union.

A3 (*Associativity*) For a composite cobordism
$Y = Y_1 \cup_{\Sigma_2} Y_2$

$Z(Y) = Z(Y_2)Z(Y_1) \in \text{Hom}(Z(\Sigma_1), Z(\Sigma_3))$.

[Note: In this associative axiom we have used the previous
two axioms to view $Z(Y_1)$ and $Z(Y_2)$ as homomorphisms
$Z(\Sigma_1) \to Z(\Sigma_2)$ and $Z(\Sigma_2) \to Z(\Sigma_3)$ respectively.]

In addition we impose the non-triviality axioms.

A4 $Z(\varnothing) = C$ for the empty d-manifold.

A5 $Z(\Sigma \times I)$ is the identity endomorphism
of $Z(\Sigma)$.

The functoriality of Z together with A5 imply *homotopy
invariance*. This means that the group $\text{Diff}^+(\Sigma)$ of orientation-
preserving diffeomorphisms of Σ acts on $Z(\Sigma)$ via its *group
of components* $\Gamma(\Sigma)$.

For a closed $(d+1)$-dimensional manifold Y the boundary
is empty and so, by A4, the vector $Z(Y)$ is just a complex
number. Thus such a topological QFT assigns numerical
invariants to closed $(d+1)$-dimensional manifolds. Moreover
cutting Y along a d-manifold Σ and applying A3 (with
$\Sigma_1 = \Sigma_3 = \varnothing$) we see that

$$Z(Y) = \langle Z(Y_1), Z(Y_2) \rangle, \qquad (2.1.1)$$

the pairing $\langle , \rangle$ being between the dual spaces $Z(\Sigma)$ and
$Z(\Sigma^*)$. Thus the numerical invariants of a closed $(d+1)$-
manifold can be computed from any decomposition $Y =
Y_1 \cup_{\Sigma} Y_2$.

The character of this representation of $\Gamma(\Sigma)$ on $Z(\Sigma)$ is
determined by the axioms. If $f \in \text{Diff}^+(\Sigma)$ we can form the

manifold Σ_f from the product $\Sigma \times I$ by using f to identify $\Sigma \times 0$ and $\Sigma \times 1$. Formula (2.1.1) then implies that

$$\text{Trace } Z(f) = Z(\Sigma_f) \qquad (2.1.2)$$

where $Z(f)$ is the induced transformation on $Z(\Sigma)$. In particular, taking f to be the identity

$$\dim Z(\Sigma) = Z(\Sigma \times S^1). \qquad (2.1.3)$$

These formulae should be compared with the Feynman integral formula (1.1.1) in the physical interpretation which we come to next.

At this stage it may be helpful to make some remarks on the physical interpretation of our axioms. The idea is that, for a closed $(d+1)$-manifold Y, the invariant $Z(Y)$ is the partition function given by some Feynman integral as discussed in Chapter 1. Of course only very special Lagrangians will give rise to topologically invariant partition functions. The vector space $Z(\Sigma)$ is then the 'Hilbert space' of the theory on the 'space' Σ. The endomorphism of $Z(\Sigma)$ given by $Z(\Sigma \times I)$ should be the 'imaginary time' evolution operator e^{-TH} (where T is the length of the interval I), but axiom A5 implies that the Hamiltonian $H = 0$. Thus, in a topological QFT there is no dynamics. All states are ground states and this is related to the finite dimensionality of the 'Hilbert space' $Z(\Sigma)$. Although there is no interesting propagation along a cylinder there is interesting propagation across a non-trivial cobordism, i.e. across singular surfaces which change the topology of Σ. This 'topological propagation' is the essential content of the theory from the Hamiltonian point of view. Relativistic invariance asserts that the final numerical invariants, such as $Z(Y)$, are independent of the time variable which one may pick to slice Y.

We are now going to concentrate on the situation that is relevant to the Jones–Witten theory. In particular we will put $d = 2$ so that Σ is a surface. In fact we need to refine and supplement the basic axioms above in a number of ways. In the first place our theory will be a *unitary* one. This means that the vector spaces $Z(\Sigma)$ all have natural Hermitian metrics (i.e. they are finite-dimensional *Hilbert spaces*) and if $\partial Y = \Sigma_2 \cup \Sigma_1^*$ the linear maps

$$Z(Y): Z(\Sigma_1) \to Z(\Sigma_2)$$

$$Z(Y^*): Z(\Sigma_2) \to Z(\Sigma_1)$$

are *adjoints* of each other. In particular, for a closed 3-manifold Y, when $Z(Y)$ is just a complex number,

$$Z(Y^*) = \overline{Z(Y)}.$$

It is this property which eventually explains the ability of the Jones polynomial to distinguish mirror images.

So far we have axiomatized an 'absolute' theory and this will lead to Witten's invariants for closed 3-manifolds. However, to get invariants for links in 3-manifolds, and hence the Jones polynomials, we have to relativize our axioms. We therefore consider a pair (Y, L) where Y is an oriented 3-manifold as before and $L \subset Y$ is an oriented 1-manifold. If Y has boundary Σ then L is assumed transversal to Σ and so $\partial L \subset \Sigma$ is an oriented 0-manifold, i.e. a collection of signed points. A typical picture is depicted below.

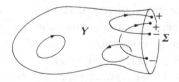

If Y is closed then L is just an oriented link in Y.

Our link L, and hence its boundary, is also assumed to carry some further information. In an abstract form this could be just an index from some given indexing set I. This is the formulation given in [29] for conformal field theories. However, for the concrete case of the Witten–Jones theory, I is just the set of irreducible representations of the compact Lie group G. Thus for each component of L (or to each signed point in Σ) we assign an irreducible representation λ of G. Reversing the orientation of the component and simultaneously replacing λ by its dual λ^* we regard as giving equivalent data.

In this framework our topological QFT is a functor Z which assigns a vector space $Z(\Sigma, P, \lambda)$ to each surface Σ with points $P = (P_1, \ldots, P_r)$ marked with representations $\lambda = (\lambda_1, \ldots, \lambda_r)$. Note that we can give each point P_i a $+$ sign by picking λ_i or λ_i^* as necessary. If Y is a 3-manifold with an oriented link L marked with representations μ, then Z assigns a vector

$$Z(Y, L, \mu) \in Z(\Sigma, \partial L, \partial \mu)$$

where $\lambda = \partial \mu$ is the induced marking on the signed set of points $P = \partial L$.

The axioms for Z have to be modified in a relatively obvious manner. Note that, for a closed 3-manifold Y with a marked link L we get a numerical invariant

$$Z(Y, L, \mu) \in C.$$

Taking $Y = S^3$, $G = SU(2)$ and μ_i the standard two-dimensional representation this invariant will eventually be identified with a certain value of the Jones polynomial. Note that the group of components of $\mathrm{Diff}^+(S^2, P)$ acts on the vector space $Z(S^2, P, \mu)$, provided all μ_i are equal. This is closely related to the braid group representations of the Jones theory. In fact B_n is the group of components of orientation-preserving diffeomorphisms of S^2 with $n + 1$ marked points $P_1, \ldots, P_{n+1}$ where $P_{n+1} = \infty$ is distinguished and kept fixed, while the others are permuted.

There is a final refinement which is necessary for the Witten theory and that concerns *framings*, i.e. trivializations of the tangent bundles of Y. These are necessary in order to pin down scalar factors in the theory. Without framings only the projective spaces of $Z(\Sigma)$ are well defined. We shall not enter into these questions but refer to Witten [36]. See also [4] for supplementary comments on framings.

2.2 Canonical quantization

Our aim is, in due course, to show how to construct the Witten functor $Z(\Sigma)$ axiomatized in the preceding section. We shall build up to this in stages, beginning first with a review of standard quantization and its application to the homology of surfaces.

Given a compact oriented surface Σ of genus g, its cohomology group $H^1(\Sigma, R)$ is a real vector space of dimension $2g$ endowed with a symplectic form ω. This is given by the cup product, followed by evaluation (integration) on the fundamental cycle. If we fix a Riemannian metric on Σ, then $H^1(\Sigma, R)$ can be identified with the space of harmonic 1-forms, and so can be thought of as a space of *classical fields* on Σ. Thus we have a functor, constructed from classical fields, which assigns a symplectic (linear) manifold to Σ and is *additive* under disjoint sums.

We can now *quantize* the symplectic manifold $H^1(\Sigma, R)$. This will produce a Hilbert space $\mathscr{H}(\Sigma)$ of *quantum* fields which will be a *multiplicative* functor, i.e.

$$\mathscr{H}(\Sigma_1 \cup \Sigma_2) \cong \mathscr{H}(\Sigma_1) \otimes \mathscr{H}(\Sigma_2)$$

when $\otimes$ denotes the completed tensor product of Hilbert spaces.

Let us review the rudiments of this quantization process. Explicitly we can construct $\mathscr{H}(\Sigma)$ as follows. First pick symplectic coordinates $p_1, \ldots, p_g, q_1, \ldots, q_g$ for $H^1(\Sigma, R)$, so that the symplectic form ω is

$$\sum_{i=1}^{g} q_i \wedge p_i.$$

Note that such coordinates arise naturally by viewing Σ explicitly as a connected sum of tori, and picking the standard basis for each torus. We now take $\mathcal{H}(\Sigma)$ to be the Hilbert space of square-integrable functions of $q_1, \ldots, q_g$. The p_j and q_j then act on $\mathcal{H}(\Sigma)$ in the well-known way: the q_j by multiplication and p_j by $i\partial/\partial q_j$. These satisfy the Heisenberg commutation relations

$$[p_j, q_k] = i\delta_{jk}.$$

Because of the Stone–von Neumann uniqueness theorem the projective space of $\mathcal{H}(\Sigma)$ is independent of the choice of symplectic coordinates. Equivalently $\mathcal{H}(\Sigma)$ is a projective representation of the symplectic group.

There is another way to construct $\mathcal{H}(\Sigma)$ involving complex coordinates which is very important and particularly relevant here. Putting $z_j = q_j + ip_j$ we can identify $\mathcal{H}(\Sigma)$ with an appropriate completion of the space of polynomials in $z_1, \ldots, z_g$.

More intrinsically we pick a complex structure on $H^1(\Sigma, R)$ so that the symplectic form comes from a Hermitian metric. There is then a holomorphic line-bundle L with connection on $H^1(\Sigma, R)$ whose curvature is the 2-form $i\omega$. The square-integrable holomorphic sections of L then give the required model of $\mathcal{H}(\Sigma)$. Again the projective space of $\mathcal{H}(\Sigma)$ is independent of the complex structure chosen.

The choice of admissible complex structure on $H^1(\Sigma, R)$ depends on a point σ of the Siegel upper half space $\mathcal{S}$ (the homogeneous space $Sp(2n, R)/U(n)$). The family $\mathcal{H}_\sigma$ of holomorphic Hilbert spaces just described forms in a natural way a bundle of Hilbert spaces over $\mathcal{S}$. Naturality means that the symplectic group $Sp(2n, R)$ acts on the bundle. The fact that the projective spaces of the $\mathcal{H}_\sigma$ can be naturally identified means that the bundle of Hilbert spaces has a natural connection which is projectively flat. Alternatively, tensoring the bundle by a suitable line-bundle over $\mathscr{S}$, we get a flat connection. The required line-bundle L is actually $K^{-1/2}$ where K is the canonical line-bundle of $\mathscr{S}$.

For a full treatment of the quantization process we refer the reader, for example, to [14] or [38].

Notice that a natural way to get a complex structure σ on $H^1(\Sigma, R)$ is to fix a complex structure τ on Σ. This enables us to identify

$$H^1(\Sigma, R) \cong H^1(\Sigma, \mathcal{O}) = H^{0,1}(\Sigma),$$

the dual of the space of holomorphic differentials on Σ. The complex structures on Σ modulo the identity component of $\mathrm{Diff}^+(\Sigma)$ are parametrized by the Teichmüller space $\mathscr{T}$ and $\tau \to \sigma$ defines an embedding $\mathscr{T} \to \mathscr{S}$, essentially the period mapping. The quantizations we are now discussing can be carried out over the whole of $\mathscr{S}$, but in the non-abelian situation to be discussed later this will no longer be true and we shall be restricted to $\mathscr{T}$.

If, in all this story, we rescale the symplectic form ω by a factor k then nothing essentially alters except that the Heisenberg commutation rules now pick up a similar factor. Physically k plays the role of the inverse of Planck's constant $\hbar$. Geometrically, however, our compact surface Σ provides a natural normalization for the 2-form ω. This normalization becomes more significant when, in the next section, we introduce the integer lattice $H^1(\Sigma, Z)$ and the associated Θ-functions. The factor k can then only take integer values and is called the *level* of the theory.

2.3   Θ-functions

We now introduce the integer lattice

$$\Lambda = H^1(\Sigma, Z) \subset H^1(\Sigma, R)$$

and correspondingly the quotient torus

$$H^1(\Sigma, U(1))$$

where we identify $U(1)$ with R/Z in the standard way by $\theta \to \exp(2\pi i \theta)$. Roughly speaking quantizing the torus should be the same as taking the Λ-invariant part of the quantization $\mathcal{H}(\Sigma)$ of the vector space $H^1(\Sigma, R)$. The different models of $\mathcal{H}(\Sigma)$ then lead to different models for the quantization of the torus.

The complex quantization is in many ways the easiest to understand and is the most relevant for our purposes. For this we pick a complex structure σ on $H^1(\Sigma, R)$, which might come from a complex structure τ on Σ itself. This makes the torus $H^1(\Sigma, U(1))$ into an abelian variety A_σ, which is the Jacobian J_τ if $\sigma = \tau$. The complex line-bundle L on $H^1(\Sigma, R)$ with curvature $2\pi i \omega$ descends to become a holomorphic line-bundle on A_σ with first Chern class represented by ω. This has 'degree' 1, in the sense that the Liouville volume

$$\int \frac{\omega^g}{g!} = 1. \tag{2.3.1}$$

The line-bundle L obtained in this way is not uniquely defined by its curvature since the torus is not simply connected. We can alter L by tensoring with any flat line-bundle. These different choices correspond to different actions of Λ on L, which we did not specify.

There are various equivalent ways to get rid of this ambiguity in L. The classical algebro-geometric way, when $\sigma = \tau$, is to consider first the degree $(g-1)$ Jacobian J_τ^{g-1}, i.e. the moduli space of holomorphic line-bundles of degree $g-1$ over Σ_τ. This has a natural divisor D given by the image of the $(g-1)$st symmetric product. This divisor, called the 'theta-divisor', represents line-bundles of degree $g-1$ on Σ_τ which have a non-zero holomorphic section. The line-bundle $[D]$

on J_τ^{g-1} defined by this divisor is then unambiguously defined. Moreover it has the correct first Chern class. To shift back to J_τ (i.e. the degree 0 Jacobian) we have to pick a base point on J_τ^{g-1}. This can be done by choosing a *spin structure on* Σ_τ or equivalently a square root of the canonical line-bundle. Having chosen such a spin structure we shift $[D]$ back to J_τ and this becomes our choice of L.

To quantize J_τ we then take the space of holomorphic sections of L. This is just one-dimensional, corresponding on J_τ^{g-1} to the section of $[D]$ vanishing on D. This also follows from the Riemann–Roch theorem using (2.3.1).

Quantizing at level k means replacing L by L^k and, again by Riemann–Roch, we get a space of dimension k^g.

The basic section of L is given transcendentally by the classical Θ-function. This is obtained by considering directly the action of Λ on the holomorphic sections of L and finding the unique fixed vector. Note that this is not strictly a vector in the Hilbert space $\mathscr{H}(\Sigma)$: it is holomorphic but not square-integrable.

More generally the sections of L^k are given by the Θ-functions of level k. If the complex structure $\sigma = \tau$ is represented in $\mathscr{S}$ by the $g \times g$ complex symmetric matrix Z (with positive definite imaginary part), and $u \in C^g$, then

$$\Theta_m(u, Z) = \sum_{\substack{l \in Z^g \\ l \equiv m \bmod k}} \exp\left[\frac{\pi i}{k} \langle l, Zl \rangle + 2\pi i \langle l, u \rangle \right]$$

(2.3.2)

is the explicit formula for the basic Θ-functions of level k. Here $m \in (Z/k)^g$ runs over the k^g basic elements and the torus A_σ is the quotient of C^g by the g basis vectors and the g columns of Z.

Although we concentrated on the case of the Jacobian, i.e. for complex structures $\sigma = \tau$, formula (2.3.2) defines Θ-functions for general σ (i.e. for general Z).

Thus the quantization of $H^1(\Sigma, U(1))$ at level k produces a vector space V_σ of dimension k^g, for each $\sigma \in \mathscr{S}$. These form a holomorphic vector bundle V over $\mathcal{S}$. As with the quantization of a linear space we expect all the projective spaces $P(V_\sigma)$ to be naturally isomorphic. If V_σ was actually a subspace of the Hilbert space $H_\sigma(\Sigma)$ (at level k) this would be automatic. Unfortunately, as was pointed out earlier, V_σ lies in some completion of H_σ. However, the explicit formulae for Θ-functions can be used to derive the necessary identifications. The result is that *the vector bundle V over $\mathcal{S}$ has a natural connection which is projectively flat*. Moreover the central (scalar) curvature can be computed explicitly.

The connection arises from the fact that the Θ_m of (2.3.2) obviously satisfy a differential equation

$$\frac{1}{k} \cdot \frac{\partial^2 \Theta_m}{\partial u_i\, \partial u_j} = 2\pi \mathrm{i}(1 + \delta_{ij}) \frac{\partial \Theta_m}{\partial Z_{ij}}.$$

This shows that the Θ_m are covariant constant sections of a connection over $\mathcal{S}$. This connection is not however totally 'natural'; it is not invariant under the action of $Sp(2g, Z)$. The natural connection differs by a central factor.

The projective flatness of the spaces $P(V_\sigma)$ can also be interpreted as a cohomological rigidity. In fact we can form the finite Heisenberg group Γ_k from the Z/k-module $H^1(\Sigma, Z/k)$ and $P(V_\sigma)$ is essentially the Heisenberg representation of Γ_k.

We now have at least the beginnings of the data needed for a topological quantum theory as described in § 2.1. We have associated a projective space to each oriented surface Σ. The next set of data would be to show how a 3-manifold Y with $\partial Y = \Sigma$ picks out a point in this projective space. Now it is not hard to see that the image of

$$H^1(Y, Z/k) \to H^1(\Sigma, Z/k)$$

is a Lagrangian sub-module W (i.e. a maximal sub-module on which the symplectic form vanishes). We could now use this Lagrangian subspace to construct the Heisenberg representation of Γ_k and W would then define a natural 'vacuum vector' in this space.

This would be the outline of the *abelian* theory where $G = U(1)$. We shall not pursue this (rather uninteresting) case in further detail. Instead we move on to study the non-abelian case, beginning in Chapter 3 with the classical theory generalizing that of the Jacobian.

3

Non-abelian moduli spaces

3.1   Moduli spaces of representations

In Chapter 2 we studied the torus $H^1(\Sigma, U(1))$ which parametrizes homomorphisms

$$\pi_1(\Sigma) \to U(1).$$

We shall now consider the space $H^1(\Sigma, G)$ which parametrizes conjugacy classes of homomorphisms

$$\pi_1(\Sigma) \to G$$

where G is any compact simply connected Lie group. For simplicity we shall frequently work with the special case $G = SU(n)$.

Now $\pi_1(\Sigma)$ has generators $A_1, \ldots, A_g, \ldots, B_1, \ldots, B_g$ with the one relation

$$\prod_{i=1}^{g} [A_i, B_i] = 1. \tag{3.1.1}$$

It follows that $H^1(\Sigma, G)$ is the quotient by G of the subset of G^{2g} lying over 1 in the map $G^g \times G^g \to G$ given by $\prod [A_i, B_i]$. This shows that $H^1(\Sigma, G)$ is a compact Hausdorff space. More precisely it is a manifold of dimension $2(g-1) \dim G$ at all *irreducible* points (i.e. where the image of $\pi_1(\Sigma)$ generates G). This follows by examining the linearization of (3.1.1). This has been examined in great detail by Narasimhan and Seshadri [22] and also by Newstead [23].

If $\alpha: \pi_1(\Sigma) \to G$ is irreducible then the tangent space to $H^1(\Sigma, G)$ at α can be identified with $H^1(\Sigma, \mathfrak{g}_\alpha)$, the cohomology of Σ with values in the flat Lie-algebra-valued bundle associated to α.

We now fix, once and for all, a G-invariant metric on the Lie algebra of G. For simple G there is only a scalar factor to be fixed and we have to choose a convenient normalization. For $SU(n)$ we take the standard metric – Trace A^2.

Using this metric and the cup product then gives a symplectic structure to $H^1(\Sigma, \mathfrak{g}_\alpha)$. In fact, as we shall see in the next section, this makes the irreducible part of $H^1(\Sigma, G)$ a *symplectic manifold*. This generalizes the symplectic structure of the torus $H^1(\Sigma, U(1))$ which we studied in Chapter 2.

Note that $g = 0, 1$ are special cases since all representations are then reducible. Where these low values of g cause problems we will usually assume $g \geq 2$.

Although we have, for simplicity, introduced $\pi_1(\Sigma)$, which requires a choice of base point, the space $H^1(\Sigma, G)$ is independent of this choice. This is because we factored out by conjugation.

It follows that the group $\text{Diff}^+(\Sigma)$ acts on $H^1(\Sigma, G)$ and it preserves the symplectic structure.

We shall discuss briefly the generalization of all this for a surface with marked points as previewed in Chapter 2.

Given a marked point P on Σ we associate to it a conjugacy class C of G of order k. Thus, if $G = SU(n)$, C is the class of a matrix with eigenvalues

$$\exp\left(\frac{2\pi i\lambda_j}{k}\right), \quad \lambda_j \text{ integral}, \sum \lambda_j = 0.$$

Given marked points $P_1, \ldots, P_r$ on Σ and associated conjugacy classes $C_1, \ldots, C_r$ we consider homomorphisms

$$\pi_1(\Sigma - (P_1 \cup P_2 \cup \cdots \cup P_r)) \to G$$

such that the loop around each P_i goes into C_i. Factoring out by conjugacy gives us a moduli space generalizing $H^1(\Sigma, G)$. We might denote this by $H^1(\Sigma, P, G, C)$.

These more general moduli spaces have been studied by Seshadri and others [30] [20]. They share the general properties of the earlier moduli spaces. In particular they are symplectic manifolds with singularities. For example if $\Sigma = S^2$ then $\pi_1(\Sigma - (P_1 \cup \cdots \cup P_r))$ is free on $r-1$ generators, and the moduli space is the quotient by G of the fibre over 1 in the multiplication map

$$C_1 \times C_2 \times \cdots \times C_r \to G.$$

The dimension of the generalized moduli space is, in general, given by the formula

$$2(g-1) \dim G + \sum_j \dim C_j.$$

The symplectic structure is, as we shall see in § 3.2, partly derived from that of $H^1(\Sigma, G)$ and partly derived from the symplectic structures of the homogeneous spaces C_j.

3.2   Moduli spaces of holomorphic bundles

In the abelian case we have already used the classical result that $H^1(\Sigma, U(1))$ can be identified with the Jacobian of Σ_τ, once a complex structure τ has been fixed on Σ. Similar results hold in the non-abelian case. First, however, we have to describe the analogues of the Jacobian. These are the moduli spaces of holomorphic G^c bundles over Σ_τ, where G^c is the complexification of G. For simplicity we shall restrict ourselves here to the case $G = SU(n)$ and refer to [5] for the more general case.

The important notion here is that of stability of holomorphic bundles. A holomorphic vector bundle E over a Riemann surface Σ_τ is said to be *stable* if, for all holomorphic subbundles F, we have

$$\frac{\deg F}{\operatorname{rank} F} < \frac{\deg E}{\operatorname{rank} E}. \tag{3.2.1}$$

Here degree means the value of the first Chern class. For an $SL(n, C)$-bundle $c_1 = 0$ and so (3.2.1) simply amounts to deg $F < 0$. Semi-stable bundles are defined similarly by requiring deg $F \leq 0$.

A ιeorem of Narasimhan and Seshadri [22] asserts that the isomorphism classes of stable holomorphic bundles of rank n form a non-singular Zariski open set $M_s(n)$ in a projective algebraic variety $M(n)$. Moreover $M(n)$ is obtained from the semi-stable bundles by an equivalence relation (stronger than just isomorphism).

Since every flat bundle is automatically holomorphic it is not surprising that there is a natural map

$$H^1(\Sigma, SU(n)) \to M(n).$$

The main theorem of Narasimhan and Seshadri [22] is that this map is a homeomorphism. The significance of this result will become clearer in Chapter 4. At the level of tangent spaces, at an irreducible point α, it corresponds to the natural isomorphism

$$H^1(\Sigma, \mathbf{End}_0 (E)) \to H^1(\Sigma_\tau, \mathrm{End}_0 (E)) \qquad (3.2.2)$$

where End_0 denotes trace-free endomorphisms and the two sheaves are:

$$\mathbf{End}_0 (E) = \text{locally constant skew-Hermitian} \\ \text{endomorphisms}$$

$$\mathrm{End}_0 (E) = \text{holomorphic endomorphisms.}$$

This is the obvious generalization of the result used in Chapter 2 for the Jacobian. However, in that case, since the manifold is a torus, the whole map is linear. Here the manifolds are non-linear and only the linearized tangent map can be easily identified by sheaf cohomology.

From (3.2.2) it is clear that the complex structure induced on $H^1(\Sigma, SU(n))$ by a complex structure on Σ depends only on the isomorphism class of Σ (modulo the identity component of $\text{Diff}^+(\Sigma)$), i.e. on the point τ in Teichmüller space. This is the generalization of the fact that the complex structure on $H^1(\Sigma, U(1))$ depends only on the period matrix of the Riemann surface. This gives a certain rigidity to our moduli spaces, a property not shared by the family of Riemann surfaces.

The moduli space $M(n)$ has a natural holomorphic line-bundle L and the space of sections of L^k will give the quantization at level k. We shall postpone a discussion of these questions until Chapter 5 where they will appear in proper context. However, it may be worth remarking at this stage that L generates the group of holomorphic line-bundles on $M(n)$ as shown by Drezet and Narasimhan [12]. Unlike the Jacobian case there are no flat line-bundles and for this reason spin structures on Σ are not needed.

There is a generalization of stability and of the moduli space $M(n)$ to take account of marked points. This is due to Seshadri [30] and involves assigning *weights* $\alpha_1, \ldots, \alpha_r$ at each marked point. Seshadri proves that his moduli space (for given weights) is naturally homeomorphic to the space of unitary representations of $\pi_1(\Sigma - (P_1 \cup \cdots \cup P_r)$ where the loop around a marked point P is represented by a matrix with eigenvalues

$$\exp(2\pi i \alpha_j), \quad j = 1, \ldots, n.$$

This is the moduli space of representations we met in the preceding section, except that we restricted the eigenvalues to be kth roots of unity, i.e.

$$\alpha_j = \frac{\lambda_j}{k}, \quad \lambda_j \text{ integral.}$$

In addition we described the case of $SU(n)$, rather than $U(n)$.

All these moduli spaces have a natural line-bundle L_k whose holomorphic sections give the quantization at level k. Note that here the moduli space itself depends on k, whereas in the absence of marked points the moduli space is independent of k and $L_k = L^k$ is the kth power of a fixed line-bundle L.

As an example we give Seshadri's definition of stability for a rank 2 vector bundle with just one marked point P. There are just two weights (assumed distinct) and we choose them so that

$$0 \le \alpha_1 < \alpha_2 < 1.$$

We first define the degree of E, relative to these weights α, by

$$\deg_\alpha E = \deg E + \alpha_1 + \alpha_2.$$

Next we fix a line L (one-dimensional subspace) of the fibre E_P which we regard as part of the structure of E. Given any holomorphic sub-line-bundle F of E we then define

$$\deg_\alpha F = \begin{cases} \deg F + \alpha_2 & \text{if } F_P = L \\ \deg F + \alpha_1 & \text{otherwise.} \end{cases}$$

E (or better (E, L)) is then defined to be stable, relative to α, if for all F

$$\deg_\alpha F < \tfrac{1}{2} \deg_\alpha E.$$

4
Symplectic quotients

4.1   Geometric invariant theory

This chapter is in the nature of a digression to discuss the formation of quotients in algebraic geometry and its relation to corresponding notions in classical and quantum mechanics. In the next chapter we shall apply these ideas in an infinite-dimensional context in order to get a better understanding of the moduli spaces discussed in the last chapter.

We begin by reviewing classical invariant theory and its geometric interpretation as developed by Mumford [21].

If A is a polynomial algebra (over C) and G is a compact group of automorphisms then the algebra A^G of *invariants* is *finitely generated*. More generally the same applies if A is replaced by a finitely generated algebra, i.e. a quotient of a polynomial algebra.

There are graded and ungraded versions of invariant theory. Geometrically these correspond to affine and projective geometry respectively. We shall be interested in the graded projective case.

If A is the graded coordinate ring of a projective variety X then its subring of invariants A^G should be the coordinate ring of some quotient projective variety. This quotient should be approximately the space of G^c-orbits in X, where G^c is the complexification of G. However, since G^c is non-compact its orbit structure can be bad and the precise nature of the quotient construction is slightly subtle. Mumford's geometric invariant theory makes this precise, and we shall now rapidly review the main features.

Abstractly we start from a smooth projective variety X with an ample line-bundle $\mathscr{L}$, i.e. such that some power of $\mathscr{L}$ defines a projective embedding of X. We assume G (or G^c) acts on X and on $\mathscr{L}$. Mumford then defines a Zariski open set X_s of X consisting of *stable* points. The G^c-orbits in X_s are closed and the quotient space $Y_s = X_s/G^c$ is a well-defined smooth quasi-projective variety. To get a natural projective compactification Y of Y_s Mumford defines the subset X_{ss} of *semi-stable* points in X, and Y is obtained from an equivalence relation on the G^c-orbits in X_{ss}. In fact Y can be identified with the closed G^c-orbits in X_{ss}.

In the best cases stable points have a trivial isotropy group. In such cases the line-bundle $\mathscr{L}$ descends naturally to a line-bundle L on Y_s and then can be extended to Y. Almost equally good is the case when the isotropy groups are finite. Then, for a suitable integer k, $\mathscr{L}^k$ descends to give a bundle on Y_s and then on Y.

In the free case (trivial isotropy on X_s) a G^c-invariant section of $\mathscr{L}$ on X_s descends to give a section of L on Y_s. Moreover sections which extend to X correspond to sections which extend to Y.

4.2   Symplectic quotients

In classical mechanics one deals with a phase-space which is a *symplectic* manifold X. If a compact Lie group G acts symplectically on X then (under mild assumptions) there is a *moment map*

$$\mu: X \to \text{Lie } (G)^*$$

taking values in the dual of the Lie algebra. If $\xi \in \text{Lie } (G)$ then $\langle \mu(x), \xi \rangle$ is the Hamiltonian function which generates the flow given by the action of ξ on X. As the terminology suggests the moment map generalizes the classical notion of angular momentum.

The space

$$Y = \mu^{-1}(0)/G$$

is called the reduced phase-space or symplectic quotient. It is a manifold (with singularities) and inherits a natural symplectic structure. Its dimension is, in general, given by

$$\dim Y = \dim X - 2 \dim G.$$

To distinguish it from the ordinary quotient X/G (which is not symplectic) it may be denoted by $X /\!/ G$. The good case (leading to the dimension formula above) is when the generic G-orbit is free (or has only a finite isotropy group).

There are more general constructions of symplectic quotients of the form

$$Y_\lambda = \mu^{-1}(\lambda)/G$$

where λ is a G-orbit in Lie $(G)^*$. These are related not to the invariant subring but to the part that transforms according to a given irreducible representation of G. In particular, if G is abelian, different integral points λ in Lie $(G)^*$ correspond to different characters of G.

A basic example is given by taking $G = U(1)$ acting by scalar multiplication on $C^n = X$. Giving C^n the symplectic structure from the standard Hermitian metric we find

$$\mu(z) = |z|^2.$$

It follows that the symplectic quotient $\mu^{-1}(1)/U(1)$ is the complex projective space $P_{n-1}(C)$ with the symplectic structure of its standard Kähler metric.

Note that $P_{n-1}(C) = (C^n - 0)/C^*$, the natural complex quotient by the group C^* (complexification of $U(1)$). Thus $P_{n-1}(C)$ occurs both as a symplectic quotient and as a complex algebraic quotient. This is in fact a typical story as we shall explain.

Return now to the situation of the preceding section with a compact Lie group G, and its complexification G^c, acting on an algebraic variety X with its ample line-bundle $\mathscr{L}$. We can fix a G-invariant connection on $\mathscr{L}$ and its curvature will then be a type $(1, 1)$ form corresponding to a G-invariant Kähler metric on X. For example, fix a G-invariant metric on the space of sections of $\mathscr{L}^k$ with k large, to give a projective embedding of X. The Kähler metric of X defines a G-invariant symplectic structure. We can therefore form both the Mumford quotient of X by G^c and the symplectic quotient $X /\!/ G$. It is a general theorem (see [18]) that these coincide and the symplectic structure of $X /\!/ G$ is defined by a Kähler metric on the Mumford quotient. The key step in this identification is to show that every closed G^c-orbit in X_{ss} contains a G-orbit in $\mu^{-1}(0)$.

The advantage of the symplectic quotient $X /\!/ G$ is that it is obviously compact, and we do not need to worry about stable or semi-stable points. On the other hand the complex structure is not obvious and for this the Mumford quotient is needed.

4.3   Quantization

The identification between the Mumford algebro-geometric quotient and the symplectic quotient is a 'classical' one. It has a quantum counterpart relating the G-invariant algebra A^G to the quantization of the symplectic quotient, which we shall now explain.

In order to quantize a symplectic manifold X the symplectic form ω (divided by 2π) has to be *integral*, so that $i\omega$ is the curvature of a line-bundle $\mathscr{L}$. One way to quantize X is (if possible) to pick a complex Kähler structure (so that ω is the $(1, 1)$ form defined by the Kähler metric). This makes $\mathscr{L}$ into a holomorphic line-bundle (with metric) and the quantum Hilbert space $\mathcal{H}$ is then taken to be the space of square-integrable holomorphic sections of $\mathcal{L}$.

This was precisely the procedure described in Chapter 2 when $X = C^n$. Moreover we saw that the projective space $P(\mathcal{H})$ depends only on the underlying symplectic structure of X and is independent of the choice of complex structure. Equivalently the bundle of Hilbert spaces $\mathcal{H}_\sigma$, with $\sigma \in \mathcal{S}$ the Siegel upper half space, has a natural connection which is projectively flat.

In the case when X is a *compact* symplectic manifold arising from a projective variety X with an ample line-bundle $\mathcal{L}$ the holomorphic sections of $\mathcal{L}$ will define a *finite-dimensional* quantum Hilbert space $\mathcal{H}$. However, the dependence of this on the choice of complex structure on X has to be investigated. There is no guarantee that we will get a projectively flat bundle. Good cases (giving flatness) include projective spaces and more generally homogeneous symplectic manifolds, 'co-adjoint orbits', of compact Lie groups.

Given an action by a compact group G on $(X, \mathcal{L})$, preserving the complex Kähler structure, the G-invariant part of the quantum Hilbert space $\mathcal{H}$ is then (in the generically free case) the same as the quantum Hilbert space of the Mumford quotient.

If we can start from an X where the projective Hilbert space $P(\mathcal{H})$ is independent of the choice of complex structure then it will follow that the same is true for the G-invariant part. Thus, in this case, we get a well-defined quantization of the symplectic quotient $X /\!/ G$.

Decomposing the graded rings A, A^G into their homogeneous components

$$A = \bigoplus_k A_k,$$

$$A^G = \bigoplus_k A_k^G,$$

we see that A_k^G can be interpreted as the quantum Hilbert space of the symplectic quotient $X /\!\!/ G$ at level k (i.e. replacing L by L^k). Note however that the *algebra* structure of A^G is not a symplectic invariant of $X /\!\!/ G$. In fact the algebra determines the *complex* manifold structure of $X /\!\!/ G$ as the maximal ideal space.

The non-compact linear case $X = C^n$ leads of course to infinite-dimensional Hilbert spaces and possibly non-compact symplectic quotients $X /\!\!/ G$. The flatness here follows from that for C^n, by just restricting to the G-invariant part.

One important warning should be made at this stage. Although the G-invariant part of the space $\mathcal{H}$ of holomorphic sections of $\mathcal{L}$ on X can be identified with the holomorphic sections of L on $X /\!\!/ G$ the *inner product* cannot easily be seen on $X /\!\!/ G$. By definition the norm of a section s of $\mathcal{L}$ is defined by integration over X. A G-invariant section is determined by its restriction to $\mu^{-1}(0)$, since this meets the generic G^c-orbit. However, the norm involves a double-integral, first over the G^c-orbit and then over $X /\!\!/ G$. The integration over the G^c-orbit (which contributes the volume of the orbit) cannot be seen on the quotient space.

In the next chapter we shall meet an infinite-dimensional version of the story described in this chapter. This is the version required for the Jones–Witten theory and it has features which are not present in the finite-dimensional case. In particular it starts from a *linear* case (as for C^n) but the symplectic quotient is *compact*. The present chapter provides a general background and introduction to this infinite-dimensional case.

4.4   Co-adjoint orbits

We conclude this chapter with some additional remarks about the generalized symplectic quotients

$$Y_\lambda = \mu^{-1}(M_\lambda)/G$$

where M_λ is a co-adjoint orbit, i.e. a G-orbit in Lie $(G)^*$. According to a well-known general result of Kirillov these co-adjoint orbits are the homogeneous symplectic manifolds of G. They are of the form G/H where H is the centralizer of a torus in G. Moreover they have natural complex Kähler structures which for 'integral' λ are projective algebraic. Their quantizations give the irreducible representations of G and so these parametrize the set of integral orbits. If λ^* denotes the representation dual to λ then one can verify that

$$M_{\lambda^*} = -M_\lambda.$$

The moment map for M_λ is just its natural embedding in Lie $(G)^*$. Now compare the moment maps

$$\mu: X \to \text{Lie } (G)^*$$

and

$$\mu_{\lambda^*}: X \times M_{\lambda^*} \to \text{Lie } (G)^*.$$

Clearly

$$\mu_{\lambda^*}^{-1}(0) = \mu^{-1}(M_\lambda),$$

and so

$$Y_\lambda = \mu^{-1}(M_\lambda)/G = \mu_{\lambda^*}^{-1}(0)/G.$$

Thus the quantization of Y_λ should be the G-invariant part of the quantization of $X \times M_\lambda$. But this is just the λ-covariant part of the quantization $\mathcal{H}$ of X, i.e. $\text{Hom}_G (\lambda, \mathcal{H})$.

Thus, analysing the moment maps over the different integral orbits of $L(G)^*$ is the classical counterpart of decomposing the quantization of X as a G-module.

5

The infinite-dimensional case

5.1   Connections on Riemann surfaces

We now come to the infinite-dimensional case of symplectic quotients and their quantization which is relevant to the Jones–Witten theory. This case was studied, for other purposes, in [5] and we refer the reader there for other details.

Given a compact oriented surface Σ and a compact Lie group G we consider the infinite-dimensional affine space $\mathscr{A}$ of G-connections on the trivial bundle over Σ. Note that, if G is simply connected, for example $SU(n)$, all G-bundles over Σ are trivial. The space $\mathscr{A}$ has a natural symplectic structure. If $A \in \mathscr{A}$, a tangent vector at A is a Lie-algebra-valued 1-form α. Hence, for two such tangent vectors α, β, we can define the skew pairing

$$(\alpha, \beta) = \int_{\Sigma} -\mathrm{Tr}\,(\alpha \wedge \beta).$$

Here we have written the formula in the case $G = SU(n)$. In general we replace $-$Trace by the fixed G-invariant inner product on Lie G.

The group $\mathscr{G}$ of gauge transformations, i.e. the group of smooth maps $\Sigma \to G$, acts naturally on $\mathscr{A}$ preserving its symplectic structure. An elementary calculation [5; p. 587] shows that the moment map

$$\mu\colon \mathscr{A} \to \mathrm{Lie}\,(\mathscr{G})^{*}$$

is just the curvature $\mu(A) = F_A$. Note that F_A, a Lie-algebra-valued 2-form, gives a linear function on Lie ($\mathcal{G}$), the space of Lie-algebra-valued 0-forms, by using the inner product on Lie ($\mathcal{G}$) and then integration over Σ. Hence the symplectic quotient

$$\mathcal{A} /\!\!/ \mathcal{G} = \mu^{-1}(0)/\mathcal{G}$$

is the moduli space of *flat G*-connections on Σ, or equivalently the moduli space of homomorphisms (up to conjugacy) $\pi_1(\Sigma) \to G$. This was the moduli space denoted by $H^1(\Sigma, G)$ in Chapter 3.

For an irreducible homomorphism $\pi_1(\Sigma) \to G$ the only gauge automorphisms that preserve it are the constant central automorphisms arising from the centre of G.

For semi-simple G this is finite so that we are, at least formally, in the 'good' case for the G-action. An abelian factor in G causes only minor differences because of the first Chern class.

We now turn to the holomorphic view-point by fixing a complex structure τ on Σ. This induces a natural complex structure on $\mathcal{A}$ so that we have an infinite-dimensional analogue of the linear situation discussed in Chapter 2. Moreover by taking the $(0, 1)$ part d''_A of the covariant derivative d_A of a connection A we can identify $\mathcal{A}$ with the space $\mathcal{C}$ of holomorphic structures on the trivial bundle $\Sigma \times G^c$ [5; Chapter 7]. Also the complexification $\mathcal{G}^c$ of $\mathcal{G}$, given by smooth maps of $\Sigma \to G^c$, acts naturally on $\mathcal{C}$. The moduli space M_τ of holomorphic G^c-bundles over Σ_τ is the analogue of the Mumford quotient of Chapter 4. It contains as an open set the subspace parametrizing stable G^c-bundles.

In fact we can do a little better. There is actually a holomorphic line-bundle $\mathcal{L}$ with connection over $\mathcal{C}$ whose curvature is $-2\pi i$ times the Kähler form, and $\mathcal{G}^c$ acts naturally on $\mathcal{L}$ with $\mathcal{G}$ preserving its metric and connection. This line-bundle is the *Quillen* line-bundle whose fibre at $A \in \mathcal{A} = \mathcal{C}$ is

$$\mathcal{L}_A = \det H^1(\Sigma_\tau, E_A) \otimes [\det H^0(\Sigma_\tau, E_A)]^*$$

where E_A is (for $G = U(n)$ or $SU(n)$) the holomorphic rank n vector bundle defined by d_A and det denotes the highest exterior power. The metric on $\mathscr{L}_A$ is defined by regularized determinants of Laplacians and Quillen [25] proved that this gives the right curvature. For general G there is a Quillen line-bundle for each representation (pull back from $U(n)$ by $G \to U(n)$), and in particular for the adjoint representation. In general these will give powers of the desired line-bundle $\mathscr{L}$.

The constant centre of G acts trivially on $\mathscr{A}$ and its action on the fibres of $\mathscr{L}_A$ is given, for $G = SU(n)$, by the scalar action of nth roots of 1 on the sheaf cohomology of E_A. But, since the first Chern class is zero,

$$\dim H^0(\Sigma_\tau, E_A) - \dim H^1(\Sigma_\tau, E_A) = n(1-g)$$

so that the scalar actions cancel. Thus $\mathscr{G}$ acts on $(\mathscr{A}, \mathscr{L})$ through the quotient $\tilde{\mathscr{G}}$ by the constant centre. This shows that the line-bundle $\mathscr{L}$ descends (without resorting to powers) to give a line-bundle L on the moduli space M, at least on the stable part.

A more careful examination shows that the line-bundle L extends to the whole of M_τ, the essential point being that the semi-stable bundles which are identified to give a single point of M_τ differ by extensions so that the determinant line $\mathscr{L}_A$ is the same for all.

Just as for the finite-dimensional case discussed in Chapter 4, we now have a map

$$H^1(\Sigma, G) \to M_\tau$$

which we expect, by analogy, to be a homeomorphism. This is the content of the Narasimhan–Seshadri theorem [22] already mentioned in Chapter 3. There is a direct proof by Donaldson [11] which is more in the spirit of our present context.

So far we have just described the classical picture leading to moduli spaces as quotients. We now take their quantizations, which is what we are really after.

We want to consider the quantization of the symplectic space $\mathscr{A}$ and then take its $\mathscr{G}$-invariant part. We expect this to be the same as the quantization of the symplectic quotient

$$\mathscr{A} /\!\!/ \mathscr{G} = H^1(\Sigma, G).$$

To define this quantization we will pick a complex structure τ on Σ and use the Narasimhan–Seshadri theorem to replace $H^1(\Sigma, G)$ by the moduli space M_τ. Now we quantize this, at level k, by taking the space of holomorphic sections of L^k over M_τ. We expect this to be projectively independent of τ.

In the next chapter we shall discuss the various methods that can be used to establish this key result. At this stage we merely note that $\mathscr{A}$ is too infinite-dimensional to have a genuine quantization of the right kind. This is why we make the reduction to the finite-dimensional (and compact) quotients.

5.2   Marked points

As we mentioned in Chapters 2 and 3 it is necessary for the Jones–Witten theory to generalize to the case of surfaces Σ with marked points. The situation of the previous section has a natural generalization as follows.

Given the moment map

$$\mu: \mathscr{A} \to \mathrm{Lie}\,(\mathscr{G})^*$$

defined by the curvature we can pick other orbits than zero in $\mathrm{Lie}\,(\mathscr{G})^*$. In particular, given a point P on Σ we have an evaluation $e_P: \mathscr{G} \to G$ and hence, dually, an embedding

$$\delta_P: \mathrm{Lie}\,(G)^* \to \mathrm{Lie}\,(\mathscr{G})^*.$$

The image consists of delta functions on P with values in $L(G)^*$. In particular a G-orbit M_λ in $L(G)^*$ defines a G-orbit $\delta_P(M_\lambda)$ in $\mathrm{Lie}\,(\mathscr{G})^*$.

Now fix points $P_1, \ldots, P_r$ on Σ and integral orbits (or G-representations) $\lambda_1, \ldots, \lambda_r$. This defines the G-orbit

$$M(\boldsymbol{P}, \boldsymbol{\lambda}) = \sum \delta_P(M_\lambda) \subset \text{Lie}\,(\mathcal{G})^*.$$

We can therefore, for each integer k, look at the generalized symplectic quotient

$$[\mu^{-1}(M(\boldsymbol{P}, \boldsymbol{\lambda})/k)]/\!/\mathcal{G}. \qquad (5.2.1)$$

This consists of connections which are flat outside the P_j and have appropriate δ-function curvatures at the P_j. A local model for a connection near P_j, with P_j as origin of polar coordinates (r, θ), is $A_j\,d\theta$ where A_j is in the conjugacy class of the G-orbit $(1/k)M_{\lambda_j}$ (and we identify $L(G) = L(G)^*$ using our invariant metric). The monodromy around P_j of such a connection is just $\exp\,(2\pi i A_j)$ and is a kth root of unity. Thus our symplectic quotient (5.2.1) is just the moduli space of representations which we denoted in Chapter 3 by $H^1(\Sigma, \boldsymbol{P}; G, \boldsymbol{C})$, the C_j being conjugacy classes of order k in G.

Once we pick a complex structure τ on Σ we again have the identification of this space of representations with a moduli space of holomorphic bundles. This was described in Chapter 3. Again a proof along the lines of Donaldson [11] would be most natural.

As explained in § 4.4 we can replace the generalized symplectic quotients by the usual ones. Thus consider the product

$$\mathcal{B} = k\mathcal{A} \times \prod_j M_{\lambda_j^*}$$

where $k\mathcal{A}$ stands for $\mathcal{A}$ with its symplectic form multiplied by k, or equivalently $\mathcal{L}$ replaced by $\mathcal{L}^k$. The symplectic quotient $\mathcal{B}/\!/\mathcal{G}$ can then be identified with (k times) the quotient in (5.2.1). Moreover the quantization of $\mathcal{B}/\!/\mathcal{G}$ should pick out that part of the quantization of $k\mathcal{A}$ which transforms according to the representation

$$\bigoplus_j e_{P_j}^*(\lambda_j)$$

of $\mathcal{G}$.

To summarize, for each k, we have a moduli space M_k (depending on k) with a line-bundle L_k. The space $H^0(M_k, L_k)$ is the 'multiplicity space' for the representation $\bigoplus_j e^*_{P_j}(\lambda_j)$ of $\mathscr{G}$ in the quantization at level k of the space $\mathscr{A}$.

Thus even if the quantization of $\mathscr{A}$ is not too well defined we have given a meaning to that part of it that transforms according to 'evaluation representations' of $\mathscr{G}$. Of course we still have to investigate the role of the complex structure chosen, but this question is postponed until the next chapter.

5.3   Boundary components

Working on Riemann surfaces with marked points is the natural algebro-geometric approach to the subject. Having marked points allows 'poles'. There is another approach which involves working on Riemann surfaces with boundary. This requires the use of complex analysis and associated boundary-value problems. We can pass from a surface with marked points to a surface with boundary simply by cutting out small discs around the marked points.

Each method has its own advantages. Thus surfaces with boundary can be glued together along a common boundary and this is an important operation in the theory. The analogue for marked points is to allow an algebraic curve to acquire singularities (double points). These questions will be taken up later.

Naturally the theory for surfaces with boundary requires a preliminary investigation of gauge theory on a circle. In fact this is the way conformal field theory enters the picture, and the representation theory plays an important role. We shall begin therefore by a rapid review of some basic aspects of the subject, referring for fuller details to [24].

Let S be the standard circle and let $\mathscr{A}_S$ denote the affine space of G-connections for the trivial bundle over S. The gauge group

$$\mathcal{G}_S = \text{Map}\,(S, G) = LG$$

is the loop group. It acts affinely on $\mathcal{A}_S$ with orbits of finite codimension. The orbits are determined by the monodromy of the connection around S: this is a conjugacy class in G.

LG has (for each integral $k \geq 1$) a central extension (by the circle) $\widetilde{LG}$. The co-adjoint action of LG on the dual of its Lie algebra preserves hyperplanes (codimension 1) and $\mathcal{A}_S$, together with its LG-action, can be identified with one of these. Thus the orbits of LG in $\mathcal{A}_S$ are co-adjoint orbits, and 'integral' orbits lead, by quantization, to irreducible representations of $\widetilde{LG}$ (i.e. projective representations of LG).

Now let us consider a surface Σ with boundary S (for simplicity we discuss only one boundary component, but the results are quite general). As before we consider the space $\mathcal{A}_\Sigma$ of G-connections on Σ and the group $\mathcal{G}_\Sigma$ of gauge transformations on Σ. The symplectic structure on $\mathcal{A}_\Sigma$ is defined as for closed surfaces and again we have a moment map

$$\mu_k \colon \mathcal{A}_\Sigma \to \text{Lie}\,(\mathcal{G}_\Sigma)^*.$$

This time, however, the moment map picks up a boundary term and one finds the formula

$$\mu_k(A) = F_A - A_S. \tag{5.3.1}$$

Here A_S is the restriction of A to S in the following refined sense. The restriction homomorphism

$$\mathcal{G}_\Sigma \to \mathcal{G}_S$$

actually lifts naturally to the central extension $\widetilde{LG} = \tilde{\mathcal{G}}_S$ of $LG = \mathcal{G}_S$. Passing to the Lie algebras and dualizing gives a map

$$(\text{Lie}\,\tilde{\mathcal{G}}_S)^* \to (\text{Lie}\,\mathcal{G}_\Sigma)^*.$$

Combined with the hyperplane inclusion

$$\mathcal{A}_S \to (\text{Lie}\,\tilde{\mathcal{G}}_S)^*$$

this associates to any $A \in \mathcal{A}_\Sigma$ the element of $(\text{Lie } \mathcal{G}_\Sigma)^*$ we have denoted simply by A_S.

Now pick a G-orbit W_α in $\mathcal{A}_S$, corresponding to the conjugacy class C_α of G. We can then form the generalized symplectic quotient

$$X_\alpha = \mu_k^{-1}(W_\alpha) /\!/ \mathcal{G}.$$

In view of the formula (5.3.1) this symplectic quotient can be identified with the moduli space of representations introduced in Chapter 3. It parametrizes representations of $\pi_1(\Sigma - P)$ whose monodromy around P lies in the conjugacy class C_α.

As for the case of marked points we can now restrict C_α to consist of kth roots of 1 and then quantize X_α at level k. (Note that the natural line-bundle L_k on X_α gives k times the standard symplectic form.)

As before we can, at least formally, interpret the resulting space as a 'multiplicity space'. It gives the part of the quantum Hilbert space of $k\mathcal{A}_\Sigma$ which transforms according to the representation $e_S^*(\lambda)$ of $\mathcal{G}_\Sigma$. Here $e_S \colon \mathcal{G}_\Sigma \to \tilde{\mathcal{G}}_S$ is the lift of the restriction or evaluation map and λ denotes the irreducible level k representation of $\tilde{\mathcal{G}}_S$ parametrized by the relevant orbit.

Although we shall not pursue this boundary case any farther we should point out it is, in a sense, slightly 'less singular' than the case of marked points. Analytically, taking a boundary value on a codimension one circle is less singular than evaluation at a point. This can have technical advantages.

6

Projective flatness

6.1   The direct approach

We have seen in Chapter 4 that to each complex Riemann surface Σ_τ, group G and integer k we can associate a vector space $V(\Sigma_\tau, G, k)$. We recall that this is defined as the space of holomorphic sections of the line-bundle L^k on the moduli space M_τ of holomorphic G^c-bundles on Σ_τ. The main result about these spaces is their *projective flatness* with respect to the parameter τ in Teichmüller space $\mathcal{T}$. This means that the vector spaces V_τ form a holomorphic vector bundle V over $\mathcal{T}$ and that this has a natural connection whose curvature is a scalar.

In this chapter we shall review several different approaches to this basic question. We begin in this section by describing the 'direct approach', i.e. the one which most naturally fits in with the quantization ideas we have been discussing.

The idea follows on naturally from the discussion in Chapter 4 and may be summarized as follows.

As we have seen in Chapter 5 our moduli space M is a symplectic quotient of an *infinite*-dimensional affine space. If it were the symplectic quotient of a *finite*-dimensional affine space the result would be clear. Quantizing M is just taking the invariant part of the quantization of the affine space. Since this quantization is (projectively) independent of the choice of complex structure the same follows for the invariant part.

The difficulty is therefore entirely attributable to the infinite-dimensionality of the space $\mathcal{A}$ of connections. If we write down the various formulae that express the projective independence of the quantization $\mathcal{H}$ of $\mathcal{A}$ we will find that they are obviously divergent. However, we only want to make sense out of them for the $\mathcal{G}$-invariant part of $\mathcal{H}$. Our task therefore is to consider these restricted formulae and make sense out of them by appropriate regularization.

This method has been developed by Hitchin [15] and, along slightly different lines, by Axelrod, Della Pietra and Witten [7]. Both versions have in particular to deal with the following two difficulties.

In the finite-dimensional case (and assuming the group is unimodular) the first Chern class of the complex moduli space is necessarily trivial. In the infinite-dimensional case this is not true because of an anomaly. This produces a shift in the formulae with the level k in appropriate places being replaced by $k+n$ (for $SU(n)$). We shall see this shift again in the Feynman integral calculations of Chapter 7.

The second difficulty relates to the inner product on the Hilbert spaces. The reasons for this difficulty (already present in finite dimensions) were mentioned in Chapter 4. Although unitarity is not strictly needed for the purpose of defining the Jones polynomials, it is a significant aspect of the theory, and a good proof is certainly desirable.

This 'direct proof' has of course to be generalized to include the case of surfaces with marked points. However, no essentially new features enter for this generalization. Roughly speaking the generalized moduli spaces differ from the simple moduli spaces by incorporating copies of the homogeneous symplectic manifolds of G, and these are well understood.

6.2   Conformal field theory

The second approach to the projective flatness is to fix a point P on the surface Σ and cut out a small disc D around P. The general idea is that questions on Σ can be reduced to studying the surface $\Sigma - D$ which has a boundary circle S, together with appropriate local data on the disc D. We have already seen in Chapter 5 how this brings in the representation theory of the loop group LG.

There are two key steps in developing this approach. In the first place the representations of LG admit a natural action of the group $SO(2)$ of rotations of the circle. However, closer examination shows that this action can be extended to $\text{Diff}^+(S)$, the 'Virasoro algebra' of physicists. This is all carefully explained in [24].

The next step is to observe that elements of $\text{Diff}^+(S)$ can be used to glue back the disc into Σ with a twist, thus obtaining a new complex Riemann surface. In fact all complex structures can essentially be obtained this way.

Putting these two facts together one can deduce the projective flatness.

The role of $\text{Diff}^+(S)$ in this proof becomes clearer if we note that the projective flatness of our Hilbert spaces can be reformulated as follows. Using a given complex structure τ on the Riemann surface to construct the corresponding Hilbert space (sections of L^k over the moduli space M_τ) it is clear that any automorphism of the complex structure Σ_τ will act on the Hilbert space. The projective flatness asserts essentially that $\text{Diff}^+(\Sigma)$ acts (projectively). This is just the two-dimensional counterpart of the one-dimensional version asserting that $\text{Diff}^+(S)$ acts (projectively) on the representations of the loop group.

The generalization to allow for surfaces with boundaries is quite natural in this approach and brings in no really new features.

6.3   Abelianization

The third approach to projective flatness is much more far reaching than the previous two but it has not yet been worked out in detail and remains somewhat conjectural at the present stage. Nevertheless it is potentially very important because it will in principle reduce the whole non-abelian theory to the elementary abelian case which was discussed in Chapter 2. Thus what is being proposed here is an *abelianization* procedure. This should be compared with the representation theory of compact Lie groups. As is well known this theory can, in a sense, be reduced to that of the maximal torus together with the action of the Weyl group. In our case there will also be a discrete group that plays the role of the Weyl group.

The whole programme envisaged here rests on the fundamental paper [16] of Hitchin, so we begin by reviewing this very briefly. For simplicity we shall discuss only the case $G = U(n)$ or $SU(n)$, but the extension to general G is fairly routine.

Hitchin introduces moduli spaces which generalize those introduced in Chapter 3 and which have both a holomorphic and a representation theory description. The holomorphic description starts from a complex Riemann surface Σ_τ and is concerned with 'Higgs bundles' over Σ_τ. These are pairs (V, ϕ) where V is a holomorphic rank n vector bundle and ϕ (the 'Higgs field') is a holomorphic section of $(\text{End } V) \otimes K$, where K is the canonical line-bundle of Σ_τ.

There is a natural notion of *stability* for Higgs bundles and a corresponding moduli space $\mathcal{M}$ (depending still on τ). There is a natural embedding $M \to \mathcal{M}$ given by bundles with zero Higgs field. Moreover the cotangent bundle T^*M_s of the stable points M gives a natural open set in $\mathcal{M}$, with the Higgs field being the cotangent vector. In particular this shows that $\dim \mathcal{M} = 2 \dim M$.

The characteristic polynomial of the Higgs field ϕ

$$\det (\lambda - \phi) = \lambda^n + a_1 \lambda^{n-1} + \cdots + a_n \qquad (6.3.1)$$

has coefficients $a_i \in H^0(\Sigma_\tau, K^i)$. Altogether these define a holomorphic map

$$\chi: \mathcal{M} \to \bigoplus_{i=1}^{n} H^0(\Sigma_\tau, K^i) = W$$

with the following properties:

(i) χ is proper,
(ii) the generic fibre is an abelian variety,
(iii) $\dim \mathcal{M} = 2 \dim W$,
(iv) M is an irreducible component of $\chi^{-1}(0)$.

The equation $\det(\lambda - \phi) = 0$ defines an algebraic curve

$$\hat{\Sigma}_\tau \subset T^*\Sigma_\tau$$

which is an n-fold branched covering of Σ_τ depending on a parameter $w \in W$. The fibre of χ over a generic point w of W is the Jacobian of $\hat{\Sigma}_\tau$. Note that the fibre over $w = 0$ is very degenerate and in particular M is a multiple component.

Thus our moduli space M appears in a degeneration of a family of abelian varieties. Moreover there is a natural line-bundle $\mathcal{L}$ on $\mathcal{M}$ whose restriction to M gives our standard line-bundle L over M.

Taking the sections of $\mathcal{L}^k$ over the fibres of χ we then get a vector bundle over the regular points of W, i.e. over $W - D$ where the discriminant locus D consists of $w \in W$ for which the characteristic polynomial (6.3.1) has a double root.

Sections of L^k over M can be pulled back to T^*M_s and then extended to all of $\mathcal{M}$ (provided the exceptional set has codimension ≥ 2, which holds for $g > 2$). We can therefore identify our Hilbert space $H^0(M, L^k)$ as a space of sections of the vector bundle over $W - D$.

Now this bundle has the projectively flat connection of the abelian case (given by Θ-functions as in Chapter 2). Our aim should be to identify $H^0(M, L^k)$, at least projectively, with the covariant constant sections of the bundle over $W - D$. Equivalently $H^0(M, L^k)$ should be the part of $H^0(\chi^{-1}(w), \mathcal{L}_w^k)$ left fixed by the monodromy group $\Pi = \pi_1(W - D, w)$ based at a generic point $w \in W - D$.

If established this would constitute our *abelianization*, with

Π (or rather its image in the symplectic group associated to the abelian variety) playing the role of the Weyl group.

The projective flatness as we now vary $\tau \in \mathcal{T}$ would follow as a corollary from that of the abelian case. Note that there are two types of variation of abelian variety being used here. First the branched cover $\hat{\Sigma}_\tau$ of Σ_τ varies with the branch points depending on $w \in W_\tau$. Then there is the variation of τ itself. The 'Weyl group' should therefore be the group (independent of τ) arising from the universal space

$$\mathcal{W} = \bigcup_\tau W_\tau$$

and the universal discriminant

$$\mathcal{D} = \bigcup_\tau D_\tau.$$

Hitchin's moduli space $\mathcal{M}$ has many other beautiful properties which are likely to repay further study. In particular $\mathcal{M}$ has a description (as a real manifold) by representations, namely as a moduli space of representations $\pi_1(\Sigma) \to GL(n, C)$. This generalizes the description of the moduli space M as the space of unitary representations of $\pi_1(\Sigma)$.

To deal with the case of surfaces with marked points, the notion of a Higgs bundle has to be generalized by requiring the Higgs field ϕ to have simple poles, at the marked points, with nilpotent residues.

6.4   Degeneration of curves

By whatever method, the conclusion of the preceding sections is that the Hilbert spaces Z_τ, given by $H^0(M_\tau, L^k)$, form a holomorphic vector bundle Z over Teichmüller space $\mathcal{T}$ with a projectively flat connection. Since this is natural it is acted on by Γ, the group of components of $\text{Diff}^+(\Sigma)$. If we factor out by Γ we get the moduli space

$$\mathcal{M} = \mathcal{T}/\Gamma$$

of curves of genus $g = \text{genus } \Sigma$. We cannot quite divide Z by

Γ to get a bundle over $\mathcal{M}$ because of the presence of fixed points. However, there are standard ways to rigidify curves to get around this problem and we shall ignore it. Technically Z is a Γ-equivariant vector bundle over the Γ-space $\mathcal{T}$, but it is easier to think in terms of bundles over $\mathcal{M}$.

Now $\mathcal{M}$ has a natural compactification $\bar{\mathcal{M}}$ obtained by allowing curves with double points. A key result in our theory is that the vector bundle (obtained from Z) over $\mathcal{M}$ *extends to a bundle over* $\bar{\mathcal{M}}$. This has been established by Tsuchiya, Ueno and Yamada [33] who also investigate the behaviour of the connection near the 'boundary' $\bar{\mathcal{M}} - \mathcal{M}$. Roughly, they prove that it has a simple pole (regular singular behaviour), but technically it has to be phrased in the language of D-modules.

If the abelianization programme of Hitchin can be carried through, then the extension to $\bar{\mathcal{M}}$ should follow from the abelian case by examining Θ-functions for generalized Jacobians.

The behaviour of our bundles Z at the boundary is closely related to the conformal field theory approach of gluing along boundaries of surfaces, and the resulting Verlinde algebra [34]. It would be instructive to see this correspondence examined in detail.

Recent results in this direction provide proofs that dim $H^0(M_r, L^k)$ agrees with the Verlinde formula. The relevant references are:

(1) A. Bertram and A. Szenes, Hilbert polynomials of moduli spaces of rank 2 vector bundles II. Preprint (1991).
(2) S. K. Donaldson, Gluing techniques in the cohomology of moduli space. Preprint (1992).
(3) A. Szenes, Hilbert polynomials of moduli spaces of rank 2 vector bundles I. Preprint (1991).
(4) M. Thaddeus, Conformal field theory and the cohomology of the moduli space of stable bundles. *J. Diff. Geom.* 35 (1992) 131–50.
(5) M. S. Narasimhan and T. R. Ramadas, Factorization of generalized theta functions. Preprint (1991).

7

The Feynman integral formulation

7.1 The Chern–Simons Lagrangian

So far we have presented the Jones–Witten theory from the Hamiltonian point of view. This gave functors

surface $\Sigma \to$ finite dimensional vector space $Z(\Sigma)$

3-manifold Y with $\Sigma = \partial Y$
$\to$ vector $Z(Y) \in Z(\Sigma)$

starting from the data of a compact Lie group G and an integer, k, the level.

This *Hamiltonian approach* is mathematically rigorous, although it is not yet entirely developed.

In this chapter we shall present Witten's *Feynman path-integral* approach. It is not mathematically rigorous, but it is conceptually simple, and provides a natural starting point for the theory.

Fix a compact Lie group, G, which for simplicity we take to be $SU(n)$. In the Feynman approach, one uses the *Chern–Simons Lagrangian*. Let Y be a closed oriented 3-manifold. Consider $\mathcal{A}$, the space of all G-connections on the trivial G-bundle over Y.

For any connection A its curvature F_A is a Lie-algebra-valued 2-form. In three dimensions, the dual to a 2-form is a 1-form, i.e. $*F_A$ is a 1-form. However, $\mathcal{A}$ is an affine space, and so its tangent space at any point consists of Lie-algebra-valued 1-forms.

Thus the curvature F can be viewed as a 1-form on A. Its value on a tangent vector to $\mathscr{A}$ at A is given by multiplying by F_A and integrating over Y, contracting on the Lie algebra variables. Let

$$\mathscr{G} = \text{group of gauge transformations}$$

$$= \text{Map}\,(Y, G).$$

Clearly $\mathscr{G}$ acts on $\mathscr{A}$; and F is $\mathscr{G}$-invariant. Moreover in the fibration (with singularities)

$$
\begin{array}{c}
\mathscr{A} \\
\Big\downarrow {\scriptstyle \mathscr{G}} \\
\mathscr{A}/\mathscr{G}
\end{array}
$$

F vanishes in the vertical (fibre) direction, and thus comes from the base. So F is a well-defined 1-form on $\mathscr{A}/\mathscr{G}$.

It turns out that F is a closed 1-form. Thus one would expect that F can be expressed in the form

$$F = df$$

for some function f, where f is a $\mathscr{G}$-invariant scalar-valued function on $\mathscr{A}$ determined up to a constant. One can fix this constant by requiring that

$$f = 0$$

on the trivial connection.

This works if $\mathscr{A}/\mathscr{G}$ is simply connected. Otherwise, one can only expect f to be locally defined and, globally, it will be multi-valued. In fact, $\mathscr{A}/\mathscr{G}$ is not simply connected, and f is well defined only up to integral multiples of some constant. This f is the *Chern–Simons functional.* It is well defined modulo integers, and is $\mathscr{G}$-invariant.

Explicit formula

Define

$$L(A) = \frac{1}{4\pi} \int_Y \mathrm{Tr}\,(A \wedge dA + \tfrac{2}{3}A \wedge A \wedge A)$$

where $A \in \mathcal{A}$. Here L is a multiple of f: the notation L has been used to be consistent with Witten [36].

One now verifies that L is invariant under the subgroup

$$\mathcal{G}_0 \subseteq \mathcal{G}$$

given by the connected component of $\mathcal{G}$ containing the identity. Here $\mathcal{G}$, $\mathcal{G}_0$ differ by a copy of Z; and, under a generator of $\mathcal{G}/\mathcal{G}_0$, L is not invariant: it picks up a multiple of 2π.

Thus $e^{ikL(A)}$ is a well-defined function of A, for $k \in Z$. Witten's invariant of 3-manifolds is now defined formally as the 'partition function':

$$Z(Y) = \int_{\mathcal{A}} \exp\,(ikL(A))\,\mathcal{D}A.$$

This is a very elegant definition provided one believes that the integral makes sense! More generally, we consider a closed oriented curve

$$C \subseteq Y$$

and fix an irreducible representation, λ, of G, in addition to the data required previously: G, k.

A connection A on Y then defines a parallel transport along any curve in Y. In particular, around C, one obtains a monodromy element $\mathrm{Mon}_C\,(A)$. Then

$$\mathrm{Tr}_\lambda\,\mathrm{Mon}_C\,(A) \equiv W_C(A),$$

evaluated by taking the trace in the representation λ. Here $W_C(A)$ is known as a *Wilson line*. Define

$$Z(Y, C) = \int_{\mathcal{A}} \exp\,(ikL(A)) \cdot W_C(A)\mathcal{D}A.$$

This is a generalization of $Z(Y)$. In physicists' language,

$$Z(Y) = \langle 1 \rangle$$

$$Z(Y, C) = \langle W_C(A) \rangle$$

where $\langle , \rangle$ denotes the (unnormalized) expectation value.

Of course, one can similarly deal with several components $C_1, \ldots, C_r$, to each one associating a different irreducible representation of G. Then

$$Z(Y, C_1, \ldots, C_r) = \langle W_{C_1}(A) W_{C_2}(A) \cdots W_{C_r}(A) \rangle.$$

It is important to notice that the above definitions involve no metrics or volumes. This is an indication that we have defined topological invariants.

7.2 Stationary-phase approximations

To see if the above definitions make any sense, we first of all consider the *stationary-phase approximation* $k \to \infty$. One should think of the parameter k as something like $1/\hbar$ where $\hbar$ is Planck's constant. The classical limit comes from $\hbar \to 0$.

For the rest of this section we shall only be concerned with $Z(Y)$: the generalizations $Z(Y, C_1, \ldots, C_r)$ are similar, and only slightly more complicated.

In the stationary-phase approximation, the dominant part comes from the stationary points of the exponent. That is, at points where

$$dL = 0$$

i.e. $F_A = 0$, by definition of L,

i.e. A is a flat connection and thus corresponds to a representation of $\pi_1(Y)$:

$$\alpha : \pi_1(Y) \to G.$$

Then, the stationary-phase approximation to $Z(Y)$ gives a sum of contributions, one from each of the representations α. Thus we need only look at the integral for $Z(Y)$ locally. Suppose α is a flat connection, and

$$A = \alpha + \beta$$

where β is 'small'. Then

$$L(A) = L(\alpha) + \frac{1}{4\pi} \int_Y \mathrm{Tr}(\beta \wedge \mathrm{d}_\alpha \beta) + \text{cubic terms.}$$

There are no linear terms, since $\mathrm{d}\mathscr{L} = 0$ at α. Here, $\mathrm{d}_\alpha\beta$ is the covariant derivative of β with respect to the connection α.

Define $Q(\beta) = (1/4\pi) \int_Y \mathrm{Tr}\,(\beta \wedge \mathrm{d}_\alpha\beta)$. This is the quadratic term in the expansion of $L(A)$ above. One can think of Q as a quadratic form in an infinite number of variables. Here:

$$Q(\beta) = -\frac{1}{4\pi} \langle \beta, {*}\mathrm{d}_\alpha\beta \rangle$$

where $\langle , \rangle$ is the inner product on Lie-algebra-valued 1-forms:

$$\langle \alpha, \beta \rangle = -\mathrm{Tr} \int \alpha \wedge {*}\beta.$$

Thus Q is given by a self-adjoint operator, $-{*}\mathrm{d}_\alpha$. This Q is related to the de Rham complex with respect to the coefficient system given by α. Let g_α be the flat G-bundle on Y given by the connection α, with fibre the Lie algebra g. Then we have a de Rham complex:

$$\Omega^0_\alpha \xrightarrow{\mathrm{d}_\alpha} \Omega^1_\alpha \xrightarrow{\mathrm{d}_\alpha} \Omega^2_\alpha \xrightarrow{\mathrm{d}_\alpha} \Omega^3_\alpha.$$

Since α is flat, $\mathrm{d}^2_\alpha = 0$.

We shall assume for simplicity that α is a non-degenerate representation; i.e. that the above complex has no cohomology:

$$H^*(Y, \mathrm{g}_\alpha) = 0.$$

Since H^3, H^2 are dual to H^0, H^1, these conditions essentially reduce to

$$H^0 = 0, \quad H^1 = 0.$$

Here $H^0(Y, \mathfrak{g}_\alpha) = 0$ corresponds to α being an irreducible representation and $H^1(Y, \mathfrak{g}_\alpha) = 0$ corresponds to this representation being isolated (since $\dim H^1$ is essentially the number of deformation parameters of the representation).

In this case, Q is degenerate on $d\Omega^0_\alpha$, since $*d_\alpha$ vanishes on the image of

$$d_\alpha : \Omega^0_\alpha \to \Omega^1_\alpha.$$

This corresponds to the fact that f is invariant under $\mathscr{G}$; $d\Omega^0_\alpha$ corresponds to infinitesimal gauge transformations. Factoring out $d\Omega^0_\alpha$, we find that Q is non-degenerate on $\Omega^1/d\Omega^0_\alpha$.

Let us now digress to discuss classical Gaussian integrals. We start with the one-dimensional integral:

$$\int_{-\infty}^{\infty} \frac{dx}{\sqrt{\pi}} \exp(-\mu x^2) = \frac{1}{\sqrt{\mu}}.$$

Continuing analytically, and putting $\mu = -i\lambda$, we obtain

$$\int_{-\infty}^{\infty} \frac{dx}{\sqrt{\pi}} \exp(i\lambda x^2) = |\lambda|^{-1/2} \exp\left(\frac{\pi i}{4} \operatorname{sgn} \lambda\right).$$

The n-dimensional form of this is as follows. Suppose Q is a non-degenerate quadratic form in $x_1, \ldots, x_n$. Then

$$\int \exp(iQ(x)) \frac{dx}{\pi^{n/2}} = |\det Q|^{-1/2} \exp\left(\frac{\pi i}{4} \operatorname{sgn} Q\right).$$

This holds for non-degenerate quadratic forms only (no zero eigenvalues).

Suppose we have the action of a compact group G (e.g. S^1) on a Euclidean space, X, and $Q(x)$ is a G-invariant quadratic form. Take a transversal slice of the space for the

G action. We must take into account some form of Jacobian: in fact, the appropriate quantity is the volume of an orbit.

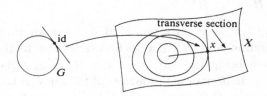

Now G acts on X. At a point $x \in X$ we have a map:

$$G \to X$$

$$g \to g(x).$$

This gives a map, B, from the tangent space of G at the identity to the tangent space of the orbit at x. The Jacobian of the map corresponds to the volume of the orbit. Thus

$$(\det B^* B)^{1/2} = \frac{\text{vol. of orbit}}{\text{vol. of } G}$$

is the appropriate scaling factor.

Hence we obtain the modulus

$$\frac{(\det B^* B)^{1/2}}{|(\det Q)|^{1/2}} \qquad (i)$$

and

$$\text{phase factor } \exp\left(\frac{\pi i}{4} \operatorname{sgn} Q\right). \qquad (ii)$$

Application to our situation

In our case,

$B = $ (infinitesimal map from Lie algebra
to tangent space of manifold)

$$= (d_\alpha : \Omega_\alpha^0 \to \Omega_\alpha^1)$$

$$\Rightarrow B^*B = d_\alpha^* d_\alpha = \Delta_\alpha^0,$$

the Laplace operator on Ω_α^0.

Now Q is given by $-*d_\alpha$ on $\Omega^1/d\Omega^0$. Consider the self-adjoint operator

$$P = \varepsilon(d_\alpha* + *d_\alpha)$$

acting on odd forms $\Omega_\alpha^{\text{odd}}$ where $\varepsilon = -1$ on Ω_α^1 and $\varepsilon = +1$ on Ω_α^3. By duality one can replace Ω_α^3 by Ω_α^0, and thus P can be thought of as acting on $\Omega_\alpha^0 \oplus \Omega_\alpha^1$. P is closely related to Q.

We can think of $\Omega_\alpha^1 = V \oplus W$ where

$$V = \text{Im}\,(d_\alpha : \Omega_\alpha^0 \to \Omega_\alpha^1)$$

and $W = V^\perp$ in Ω_α^1. Then Q acts on W; and P acts on $\Omega_\alpha^0 \oplus V \oplus W$ by

$$\begin{bmatrix} 0 & -B^* & 0 \\ -B & 0 & 0 \\ 0 & 0 & Q \end{bmatrix}$$

Ω_α^0 $\xrightarrow{d_\alpha}$ Ω_α^1 Q acts here

$\text{Im}\,d_\alpha = V$ $(\text{Im}\,d_\alpha)^\perp = W$

A quadratic form of the type

$$\begin{bmatrix} 0 & -B^* \\ -B & 0 \end{bmatrix}$$

always has zero signature. Of course, in this case, we have not assigned meanings to det Q or sgn Q; but in any 'sensible' definition one would hope that

$$\text{sgn } P = \text{sgn } Q,$$

$$|\det P| = (\det B^*B)|\det Q|. \tag{iii}$$

So, if we can make sense of det P, det B^*B, then we can write down the local contribution to the stationary phase approximation.

Here we have so far left out the level k. In finite dimensions, such a factor changes the resultant integral by an appropriate power of k, which in our case is zero.

Regularization of determinants and signatures

Let Δ be a Laplace operator, with positive eigenvalues λ. Then we can define the zeta function:

$$\text{Tr } \Delta^{-s} = \sum_\lambda \lambda^{-s} = \zeta(s).$$

The function $\zeta(s)$ is a meromorphic function, defined in the first instance for Re (s) sufficiently large. It can be analytically continued to the whole complex plane, having isolated poles. Here $s = 0$ is not a pole, and $\zeta(0)$, $\zeta'(0)$ are well defined.

Formally, $\zeta(0)$ is the dimension of the Hilbert space. In odd dimensions, $\zeta(0) = 0$.

Following Ray and Singer [26] we define

$$\det \Delta = \exp(-\zeta'(0)).$$

Formally, one sees that

$$\zeta'(0) = \sum_\lambda d/ds(\lambda^{-s})|_{s=0}$$

$$= \sum_\lambda (-\log \lambda)$$

and thus $\exp(-\zeta'(0)) = \prod_\lambda (\lambda) = \det \Delta$.

The above definition makes sense as a real number, and is used by physicists to make sense of Gaussians occurring in QFT. We wish to do this for the Laplacian with twisted coefficients Δ_α^0. This makes $\det(\Delta_\alpha^0)$ well defined, and thus gives the B^*B term.

Similarly, $P^2 = \Delta^0 \oplus \Delta^1$, the direct sum of the Laplace operators on Ω^0, Ω^1. Thus

$$(\det P)^2 = (\det \Delta_\alpha^0) \cdot (\det \Delta_\alpha^1)$$

and hence $|\det P|$ is well defined, giving $|\det Q|$, from (iii).

Thus one can evaluate (i), obtaining

$$\frac{(\det \Delta_\alpha^0)^{3/4}}{(\det \Delta_\alpha^1)^{1/4}}.$$

Ray and Singer [26] proved that

$$T_\alpha = \frac{(\det \Delta_\alpha^0)^{3/2}}{(\det \Delta_\alpha)^{1/2}}$$

the square of the above expression is independent of Riemannian metric. The Riemannian metric is used to obtain a *-operator, which is necessary to make sense of the divergent quantities.

To prove independence of metric, one differentiates T_α with respect to the metric as parameter, and shows that this vanishes. Ray and Singer conjectured that T_α was the classical Reidemeister torsion. This conjecture was proved (independently) by Cheeger and Müller. This is the first concrete encouragement for the Witten formula for $Z(Y)$: the absolute value of the limit $k \to \infty$ can be regularized, and the result is metric independent. This observation relating Ray–Singer torsion to the abelian Chern–Simons theory was made by Schwarz [28] in the late 1970s (and it extends fairly easily to the non-abelian case).

Phase factor

We now consider the phase factor as given by (ii). This involves sgn Q, which is related to sgn P, and was studied by Atiyah, Patodi and Singer [6].

Consider the situation where P is a self-adjoint operator with both positive and negative eigenvalues, and

$$\Delta = P^2.$$

Define

$$\eta(s) = \sum_{\lambda \neq 0} (|\lambda|^{-s} \operatorname{sgn} \lambda).$$

Once again, η can be analytically continued, and $\eta(0)$ is well defined. Formally,

$$\eta(0) = (\text{No. of positive eigenvalues})$$
$$- (\text{No. of negative eigenvalues})$$

and it is thus natural to define sgn $P = \eta(0)$. Note however that this quantity is a real number, not an integer. Thus the resulting phase in (ii) will not be a root of unity in general.

Then we have (cf. Proposition 4.20 of [6])

$$\operatorname{sgn} Q = \eta_\alpha(0)$$

where η_α is the η-function associated with P_α. We now have to investigate how sgn Q_α depends on the metric.

Here α is a representation of $\pi_1(M)$, with no cohomology. Consider the trivial representation, and put

$$\tilde{\eta}_\alpha(0) = \eta_\alpha(0) - \eta_d(0)$$

where $\eta_d = d\eta_1$ and η_1 corresponds to ordinary differential forms, without group fibres; d is the dimension of our Lie group. Then:

(1) $\tilde{\eta}_\alpha$ is independent of metric,
(2) $\tilde{\eta}_\alpha(0) = (4/\pi)\delta(G)L(\alpha)$,

where $\delta(G)$ is a numerical invariant of G (it is n for $SU(n)$: in general it depends on the value of the Casimir in the adjoint representation) and L is the Chern–Simons functional.

Thus we obtain from the stationary-phase formula:

$$\sum_{\alpha} (\text{contributions at } \alpha) = C\left(\tilde{\sum_{\alpha}}\right)$$

where C is a fixed multiplier, coming from η_d and C contains the only metric dependence in the formula; $\tilde{\sum}_{\alpha}$ is metric independent. The phase factor is independent of G and the chosen representation, but depends on the choice of ground metric. The above formula for $\tilde{\eta}_{\alpha}(0)$ leads to a shift

$$e^{ikL(\alpha)} \to e^{i(k+\delta)L(\alpha)}$$

in the exponential multiplier arising from the value of the action at the critical point α. Such a shift is well known to physicists in various guises.

If we had succeeded in making an expression for $Z(Y)$ independent of metric, we would have shown, for large k, how to make sense of the determinants and signatures by regularizing. This is very nearly true, but not quite – we have a phase ambiguity.

For this reason Witten has to choose a framing of the 3-manifold. For related reasons to define the invariants for links we have to choose (normal) framings for each component of the link. We shall say more about these framings in the next section.

7.3   The Hamiltonian formulation

We shall now indicate why the Chern–Simons Lagrangian is supposed to lead to the Hamiltonian version of the theory which we have been developing in earlier chapters.

To go from the path-integral to the Hamiltonian formulation we have to separate out space and time. We therefore consider a 3-manifold of the form $\Sigma \times R$. To get the Hilbert space of the theory we are supposed to quantize the space of 'classical solutions', i.e. critical points of the Lagrangian. But these are just gauge equivalence classes of flat connections and so give us the moduli space of flat G-bundles on $\Sigma \times R$. However, these are the same as the flat G-bundles on Σ. These are the moduli spaces we met in Chapter 3 whose quantizations give the (finite-dimensional) Hilbert spaces we have been studying.

The fact that these spaces are independent of the R-variable (time) shows that the Hamiltonian of the theory is zero (i.e. that the theory has no dynamics and is topological).

An alternative derivation is to re-interpret a connection A on $\Sigma \times R$ as a path A_t of connections on Σ. This comes by using parallel transport in the time direction (R) to identify bundles at different times or, as physicists would say, by working in a gauge in which $A_0 = 0$. This simplifies the Chern–Simons Lagrangian since the cubic term now drops out and we simply get

$$L = \frac{k}{8\pi} \int dt \int_{\Sigma} \operatorname{Tr} A \wedge dA.$$

This is just the classical formula for the action for a path in the symplectic linear space $\mathcal{A}$. It follows that our Hilbert space should be the $\mathcal{G}_{\Sigma}$-invariant part of the quantum Hilbert space of $\mathcal{A}_{\Sigma}$. However, as we have argued formally in Chapters 4 and 5, this should be the same as quantizing the symplectic quotient $\mathcal{A}_{\Sigma} /\!\!/ \mathcal{G}_{\Sigma}$, i.e. the moduli space of flat G-bundles over Σ.

We shall conclude this chapter with a few brief comments on the relation between the phase subtleties in the Lagrangian and Hamiltonian approaches. We recall from § 7.2 that (for the limit $k \to \infty$) there was a non-topological term, depending on a background metric. Witten shows in [36] that, by subtracting a 'counter-term' (the gravitational Chern–Simons), we can recover a purely topological theory. However, for this one has to pick a framing of the 3-manifold (itself a piece of topological data).

In the Hamiltonian version the corresponding difficulty has to do with the phase ambiguity in our Hilbert spaces: the fact that the curvature of the bundle of Hilbert spaces (over Teichmüller space) is a *non-zero scalar*.

The relation between these two manifestations of the phase ambiguity depends on earlier ideas of Witten [37], subsequently given rigorous formulation and proof by Bismut and Freed [8]. This relates the gravitational η-invariant of the 3-manifold Σ_f, constructed from $f \in \text{Diff}^+ (\Sigma)$, with the monodromy of the Quillen determinant line-bundle. We have essentially been ignoring these subtleties so it would make little sense to enter now into an elaborate discussion. We should emphasize, however, that they are a crucial aspect of the theory (related also to the central extensions of loop groups) and refer the reader to the reference above as well as [36]; see also [4].

8

Final comments

8.1 Vacuum vectors

In this final chapter we shall deal rather briefly with other aspects of the Jones–Witten theory. First of all we want to discuss how the functional integral, at least formally, gives the extra data required for a topological quantum field theory, as axiomatized in Chapter 2.

For a 3-manifold Y with boundary Σ the Chern–Simons functional $L(A)$ of Chapter 7 is not really a complex number (modulo $2\pi Z$). Intrinsically the exponential $e^{iL(A)}$ should be viewed as a vector in the complex line $\mathcal{L}_{A_\Sigma}$, the fibre of the standard line-bundle $\mathcal{L}$ over the point A_Σ in the space $\mathcal{A}_\Sigma$ of connections on the boundary Σ. For the special case $Y = \Sigma \times I$ with the boundary

$$\partial Y = \Sigma_1 - \Sigma_0, \quad \Sigma_j = \Sigma \times (j)$$

this can be seen as follows.

Using parallel transport in the I-directions we can identify connections on Y with a path A_t of connections on Σ, $0 \le t \le 1$. As noted in Chapter 7 the Chern–Simons functional then becomes the classical action for paths on a symplectic manifold, and its exponential therefore gives the parallel transport (along the path A_t in $\mathcal{A}_\Sigma$) from the fibre $\mathcal{L}_0$ to the fibre $\mathcal{L}_1$. Thus

$$e^{iL(A)} \in \mathcal{L}_0^* \otimes \mathcal{L}_1 = \mathcal{L}_{A_{\partial Y}}$$

and, raising to the kth power,

$$e^{ikL(A)} \in \mathcal{L}_{A_{\partial Y}}^k.$$

We shall now show formally how a 3-manifold Y with $\partial Y = \Sigma$ gives rise to a vector

$$Z(Y) \in Z(\Sigma)$$

in the Hilbert space $Z(\Sigma)$, as required by the axioms of Chapter 2. Recall that $Z(\Sigma)$ is defined, at level k, by a space of sections of the line-bundle L_Σ^k, where L_Σ is the line-bundle on the symplectic quotient $\mathscr{A}_\Sigma /\!\!/ \mathscr{G}_\Sigma$. We then define $Z(Y)$ as the linear function on $Z(\Sigma^*) = Z(\Sigma)^*$ given by assigning to the section ϕ of $\mathscr{L}_\Sigma^{-k}$ the Feynman integral

$$\int e^{ikL(A)} \phi(B) \mathscr{D}A.$$

This Feynman integral is over all connections A on Y which are flat (and equal to B) on Σ. Intuitively the measure $\mathscr{D}A$ involves the symplectic measure on $\mathscr{A}^\Sigma /\!\!/ \mathscr{G}_\Sigma$ together with a measure coming from the interior.

We can make more rigorous sense of this procedure, in the large k limit, by applying stationary-phase approximation as in Chapter 7. This reduces the problem to the relevant critical points which are the flat connections on Y. For example, when Y is a 'handlebody', $H^1(Y, G)$ can be identified with a Lagrangian sub-manifold of $H^1(\Sigma, G)$. For a Heegard splitting of a closed 3-manifold along a surface Σ the two Lagrangian sub-manifolds obtained from the two halves intersect at points corresponding to representations $\pi_1(Y) \to G$. This brings us back to the stationary-phase calculations made in Chapter 7, and the situation is formally similar to that of the Casson invariant which is the invariant of another topological quantum field theory (cf. [2]).

Useful recent references for the material discussed here are:

(1) T. R. Ramadas, I. M. Singer and J. Weitsman, Some comments on Chern–Simons gauge theory, *Commun. Math. Phys.* 126 (1989) 409–30.

(2) L. Jeffrey and J. Weitsman, Half density quantization of the moduli space of flat connections and Witten's semi-classical manifold invariants, *Topology* (to appear).

(1) elucidates the Chern–Simons functional, while (2) addresses the questions in the last paragraph.

8.2   Skein relations

As mentioned in Chapter 1 the Jones polynomial of links in S^3 can be characterized by a skein relation. This involves comparison of the three links obtained by various crossing changes at a fixed vertex of a planar diagram. The identification of Witten's functional integral invariant (for $G = SU(2)$ with its standard representation C^2) with a value of the Jones polynomial ($t = 2\pi i/(k+2)$)) rests therefore on demonstrating that it satisfies the same skein relation.

The fact that Witten's invariant satisfies a skein relation of the right form (for $SU(n)$ with its standard representation C^n) is an elementary consequence of the fact that the Hilbert space of Witten's theory for the 2-sphere S^2 with four marked points (two positive, two negative) has dimension 2. In fact decomposing S^3 into two balls by cutting out a small neighbourhood of the given vertex we get precisely S^2 with four marked points as common boundary. In its Hilbert space $\mathcal{H}$ we have a vector, say u, determined by the exterior, and three vectors, say v_+, v_-, v_0, determined by the three interiors (depending on the links L_+, L_-, L_0). The Witten invariants for these three links are then the scalar products in $\mathcal{H}$,

$$\langle u, v_+ \rangle, \quad \langle u, v_- \rangle, \quad \langle u, v_0 \rangle.$$

If dim $\mathcal{H} = 2$ the three vectors v_+, v_-, v_0 must satisfy a linear relation and their scalar products with u then satisfy the same relation. Note that $\mathcal{H}$ and the three vectors v_+, v_-, v_0 are locally determined and are independent of the rest of the link. Thus the coefficients of the linear relation are universal (depending only on n and k).

The reason why dim $\mathscr{H} = 2$ is the following. Quite generally the Hilbert space for S^2 with points P_i $(i = 1, \ldots, r)$ marked by representations λ_i of G can (from its definition) be shown to be a subspace of the G-invariant part of the tensor product

$$\lambda_1 \otimes \cdots \otimes \lambda_r.$$

For large k we always get the whole space. In particular take $r = 4$ and $\lambda_1 = \lambda_2 = \lambda_3^* = \lambda_4^*$. Then the dimension of the G-invariant part of

$$\lambda_1 \otimes \lambda_1 \otimes \lambda_1^* \otimes \lambda_1^* = \text{End} \, (\lambda_1 \otimes \lambda_1)$$

is the number of irreducible summands in $\lambda_1 \otimes \lambda_1$. For $G = SU(n)$ and $\lambda_1 = C^n$ this number is 2.

The computation of coefficients of the skein relation (i.e. the dependence of n and k) is given by Witten [36]. It depends on algebraic results of Verlinde [34] which Witten reinterprets in terms of surgery formulae. These ideas will be discussed briefly in the next section.

8.3   Surgery formula

If we want to compute Witten's invariant for a 3-manifold (without links) then we can proceed by using surgery. This means we consider cutting a tube $S^1 \times D^2$ out of the manifold, twisting its boundary (the torus $S^1 \times S^1$), and then inserting it back into the 3-manifold. Every 3-manifold can be obtained, starting from the 3-sphere, by a sequence of such surgeries.

The essential step in computing Witten's invariant by surgery is then to know:

(i)   the Hilbert space of a torus,
(ii) the action of the modular group $SL(2, Z)$ (the group of components of $\text{Diff}^+ \, (S^1 \times S^1)$) on this Hilbert space.

The Hilbert space of a torus can be computed in various ways. Since the fundamental group is abelian the moduli space of representations is easily determined. Thus for $SU(n)$ the moduli space is the complex projective space $P_{n-1}(C)$ with its standard line-bundle. Hence the Hilbert space can be identified with the space of homogeneous polynomials in n variables of degree k. For $n = 2$ this gives a $(k + 1)$-dimensional space.

From the point of view of loop groups the Hilbert space of a torus can be identified with the representations of LG of level k. Moreover this identification is natural once we pick an interior (solid torus). This gives an explicit basis for the Hilbert space. The action of $SL(2, Z)$ can be computed by using the Verlinde algebra. The essential point is to compute (in the explicit basis) the matrix S representing the element $\begin{bmatrix} 0 & 1 \\ -1 & 0 \end{bmatrix}$ of $SL(2, Z)$.

In principle it is also possible to compute the action of $SL(2, Z)$ from the holomorphic quantization point of view. Since π_1 is abelian we only need to know the way Θ-functions vary with the modulus of an elliptic curve as explained in Chapter 2.

8.4   Outstanding problems

Since Witten's theory involves the heuristic Feynman integral, it may be helpful to review here how much of the theory is on a rigorous basis, and what outstanding problems remain.

As we have more or less indicated the construction of the vector spaces $Z(\Sigma)$ associated to framed surfaces Σ (with or without marked points) can be done quite rigorously. The difficult part of the theory is to construct the vectors $Z(Y) \in Z(\Sigma)$ associated to framed 3-manifolds Y with boundary Σ. However, the axioms in Chapter 2 give rules governing these vectors. These rules can be used to evaluate them. The only difficulty is that the rules might not be consistent. One has therefore to check consistency.

For the Jones polynomials this was essentially the original approach of Jones. For the new invariants of 3-manifolds consistency has been verified by Reshetikhin and Turaev [27]. An alternative approach has been given in R. Kirby and P. Melvin, The three-manifold invariants of Witten and Reshetikhin for sl(2.C), *Invent. Math.* 105 (1991) 473–545.

Essentially the consistency of the Witten axioms (or rules) involves understanding how the Hilbert spaces $Z(\Sigma)$ change as a surface Σ acquires a double point. The formulation of Tsuchiya and Yamada in terms of the compactified moduli space $\overline{\mathcal{M}g}$ appears to incorporate the relevant properties, but it would be desirable to elucidate the situation. See also [19].

Defining the vectors $Z(Y) \in Z(\Sigma)$ is, as we explained in § 8.1, equivalent to computing certain Feynman integrals. Since the Chern–Simons Lagrangian is purely topological there are no real local difficulties of analytical nature in the Feynman integral. We can therefore view the surgery methods indicated above as an effective way of computing our Feynman integral. After all, an integral is simply a linear functional with certain additive local properties, and the consistency verification we have alluded to could be construed as checking these properties.

It would of course be even better if one could define some purely combinatorial version of the Chern–Simons Lagrangian as in lattice-gauge theories. Some encouragement comes from the fact that Reidemeister torsion has such a definition and this enters into the stationary-phase calculation for the Chern–Simons Lagrangian described in Chapter 7. However, this may be too ambitious and we may have to settle for the surgery approach.

In addition to the Hamiltonian approach using the Hilbert space $Z(\Sigma)$ the stationary-phase calculations of Chapter 7 also lead to rigorous formulae. Although we only gave the leading term it should be possible to proceed further and develop a fully rigorous series expansion in k^{-1}. It is then a challenging problem to show that this does in fact give the expansion of the Witten invariant computed by Hamiltonian methods. As yet this problem is very much open. Moreover, it is not clear what kind of function of k we get in general from Witten's theory. For links in S^3 the Jones invariant is a polynomial in $t = \exp(2\pi i/(k+2))$, but for general 3-manifolds the situation is more complicated. In particular it is not obvious that the Witten invariant will always be determined by its k^{-1} expansion.

There has been considerable recent progress in understanding perturbative Chern–Simons theory. In particular it has been shown in (1) that the integrals associated to Feynman graphs in the perturbation expansion for a 3-manifold are *finite* and *metric independent*. Some references are:

(1) S. Axelrod and I. M. Singer, Chern–Simons perturbation theory. Preprint (1991).
(2) M. Kontsevich, Graphs, homotopical algebra and low-dimensional topology. Preprint (1992).
(3) D. Bar-Natan, Perturbative aspects of the Chern–Simons topological quantum field theory. PhD thesis (Princeton University) 1991.
(4) D. Bar-Natan, On the Vassiliev knot invariants. Preprint (1992).

8.5   Disconnected Lie groups

In describing Witten's theory we restricted ourselves to compact connected and simply connected Lie groups G. The simple connectivity can be dropped without great difficulty. The main new feature is that there will now be topologically non-trivial G-bundles over a surface Σ.

Remarkably it is also possible to allow disconnected groups in an interesting way. In particular we can even take G to be a *finite* group. In this case the theory becomes totally rigorous from all points of view (albeit a little dull and less deep than the original Jones theory). This theory has been worked out by Dijkgraaf and Witten [9] and is of interest in physics in relation to orbifolds (quotients of manifolds by finite groups). The key point is that the level k (or rather the inner product in the Lie algebra, multiplied by k) should be interpreted as an element in $H^4(BG, Z)$, where BG is the classifying space of G. For a finite group G this is the same as $H^3(BG, R/Z)$, and a G-bundle over a closed oriented 3-manifold Y defines a map $Y \to BG$ and hence an R/Z-invariant. This is the analogue of the Chern–Simons function.

References

1. J. W. Alexander, A lemma on a system of knotted curves, *Proc. Nat. Acad. Sci. USA* 9 (1973) 93-5.
2. M. F. Atiyah, Topological quantum field theories, *Publ. Math. Inst. Hautes Etudes Sci. Paris* 68 (1989) 175-86.
3. M. F. Atiyah, New invariants of 3- and 4-dimensional manifolds, *Am. Math. Soc., Proc. Symp. Pure Maths.* 48 (1988) 285-99.
4. M. F. Atiyah, On framings of 3-manifolds, *Topology* 29 (1990) 1-8.
5. M. F. Atiyah and R. Bott, The Yang-Mills equations over Riemann surfaces, *Phil. Trans. R. Soc. Lond.* A 308 (1982) 523-615.
6. M. F. Atiyah, V. K. Patodi and I. M. Singer, Spectral asymmetry and Riemannian geometry I, *Math. Proc. Camb. Phil. Soc.* 77 (1975) 43-69.
7. S. Axelrod, S. Della Pietra and E. Witten, Geometric quantization of Chern-Simons gauge theory, *J. Diff. Geom.* 33 (1991) 787-902.
8. J. M. Bismut and D. S. Freed, The analysis of elliptic families: Dirac operators, eta invariants and the holonomy theorem of Witten, *Comm. Math. Phys.* 107 (1986) 103-63.
9. R. Dijkgraaf and E. Witten, Topological gauge theories and group cohomology, *Comm. Math. Phys.* 239 (1990) 383-429.
10. S. K. Donaldson, Polynomial invariants for smooth four-manifolds, *Topology* 29 (1990) 257-315.
11. S. K. Donaldson, A new proof of a theorem of Narasimhan and Seshadri, *J. Diff. Geom.* 18 (1983) 269-77.
12. J. M. Drezet and M. S. Narasimhan, Groupes de Picard des variétés de modules de fibrés semistables sur les courbes algébriques, *Invent. Math.* 97 (1989) 53-94.

13. A. Floer, An instanton invariant for three-manifolds, *Comm. Math. Phys.* 131 (1990) 347–80.
14. V. Guillemin and S. Sternberg, Geometric asymptotics, *Am. Math. Soc. Math. Surveys* vol 14 (1977), Providence, R.I.
15. N. J. Hitchin, Flat connections and geometric quantization, *Comm. Math. Phys.* 131 (1990) 347–80.
16. N. J. Hitchin, The self-duality equations on a Riemann surface, *Proc. Lond. Math. Soc.* 55 (3) (1987) 59–126.
17. V. F. R. Jones, Hecke algebra representations of braid groups and link polynomials, *Ann. Math.* 126 (1987) 335–88.
18. F. C. Kirwan, *Cohomology of quotients in symplectic and algebraic geometry*, Mathematical Notes 31, Princeton University Press, 1984.
19. M. Kontsevich, Rational conformal field theory and invariants of 3-dimensional manifolds (to appear).
20. V. B. Mehta and C. S. Seshadri, Moduli of vector bundles on curves with parabolic structures, *Math. Ann.* 248 (1980) 205–39.
21. D. Mumford, *Geometric invariant theory*, Springer-Verlag, Berlin, 1965.
22. M. S. Narasimhan and C. S. Seshadri, Stable and unitary vector bundles on a compact Riemann surface, *Ann. Math.* 82 (1965) 540–67.
23. P. E. Newstead, Characteristic classes of stable bundles over an algebraic curve, *Trans. Am. Math. Soc.* 169 (1972) 337–45.
24. A. Pressley and G. B. Segal, *Loop groups*, Oxford University Press, 1986.
25. D. G. Quillen, Determinants of Cauchy–Riemann operators over a Riemann surface, *Funct. Anal. Appl.* 19 (1986) 31.
26. D. B. Ray and I. M. Singer, R-torsion and the Laplacian on Riemannian manifolds, *Adv. Math.* 7 (1971) 145–210.
27. N. Yu. Reshetikhin and V. G. Turaev, Invariants of three manifolds via link polynomials and quantum groups, *Invent. Math.* 103 (1991) 547–97.
28. A. Schwarz, The partition function of degenerate quadratic functionals and Ray–Singer invariants, *Lett. Math. Phys.* 2 (1978) 247.
29. G. B. Segal, Conformal field theory, *Proceedings of International Congress of Mathematical Physics*, Swansea, 1988.
30. C. S. Seshadri, Moduli of vector bundles with parabolic structures, *Bull. Am. Math. Soc.* 83 (1977) 124–6.

31. P. G. Tait, *On knots I, II, III*, Scientific Papers, Cambridge University Press, 1900.
32. W. H. Thomson, On vortex motion, *Trans. R. Soc. Edin.* 25 (1869) 217-60.
33. A. Tsuchiya, K. Ueno and Y. Yamada, Conformal field theory on universal family of stable curves with gauge symmetry, *Adv. Stud. in Pure Maths* 19 (1989) 459-565.
34. E. Verlinde, Fusion rules and modular transformations in 2d conformal field theory, *Nucl. Phys.* B300 (1988) 360.
35. E. Witten, Topological quantum field theory, *Comm. Math. Phys.* 117 (1988) 353-86.
36. E. Witten, Quantum field theory and the Jones polynomial, *Comm. Math. Phys.* 121 (1989) 351-99.
37. E. Witten, Global gravitational anomalies, *Comm. Math. Phys.* 100 (1985) 297-9.
38. N. Woodhouse, *Geometric quantization* (second edition), Oxford University Press, 1992.

Index

Printed in the United States
By Bookmasters